Cambridge Elements

Elements in the Philosophy of Biology
edited by
Grant Ramsey
KU Leuven

BIOLOGY AND MEDICAL THEORY

Peter Takacs
Macquarie University

CAMBRIDGE
UNIVERSITY PRESS

Shaftesbury Road, Cambridge CB2 8EA, United Kingdom

One Liberty Plaza, 20th Floor, New York, NY 10006, USA

477 Williamstown Road, Port Melbourne, VIC 3207, Australia

314–321, 3rd Floor, Plot 3, Splendor Forum, Jasola District Centre, New Delhi – 110025, India

103 Penang Road, #05–06/07, Visioncrest Commercial, Singapore 238467

Cambridge University Press is part of Cambridge University Press & Assessment, a department of the University of Cambridge.

We share the University's mission to contribute to society through the pursuit of education, learning and research at the highest international levels of excellence.

www.cambridge.org
Information on this title: www.cambridge.org/9781009768061

DOI: 10.1017/9781009768047

© Peter Takacs 2025

This publication is in copyright. Subject to statutory exception and to the provisions of relevant collective licensing agreements, no reproduction of any part may take place without the written permission of Cambridge University Press & Assessment.

When citing this work, please include a reference to the DOI 10.1017/9781009768047

First published 2025

A catalogue record for this publication is available from the British Library

ISBN 978-1-009-76806-1 Hardback
ISBN 978-1-009-76801-6 Paperback
ISSN 2515-1126 (online)
ISSN 2515-1118 (print)

Cambridge University Press & Assessment has no responsibility for the persistence or accuracy of URLs for external or third-party internet websites referred to in this publication and does not guarantee that any content on such websites is, or will remain, accurate or appropriate.

For EU product safety concerns, contact us at Calle de José Abascal, 56, 1°, 28003 Madrid, Spain, or email eugpsr@cambridge.org

Biology and Medical Theory

Elements in the Philosophy of Biology

DOI: 10.1017/9781009768047
First published online: November 2025

Peter Takacs
Macquarie University
Author for correspondence: Peter Takacs, peter.takacs@mq.edu.au

Abstract: The philosophy of medicine has long been concerned with the status of diseases and disorders. Are such states genuinely pathological in an objective, mind-independent sense or merely value-laden social constructs? The prevailing dialectic in this area accordingly pits normative views against non-normative, naturalistic positions. Hybrid accounts represent a better alternative to these needlessly extreme "purist" views. Hybrid accounts maintain that objective criteria for health and disease can exist independently and harmoniously alongside of normative considerations. Hybrid accounts to date have nevertheless failed to convince many. The failure is due largely to their reliance on inadequate notions of biological dysfunction. This Element attempts to redress this situation by sketching the outlines of a more sophisticated dysfunction condition. Drawing on recent advances in evolutionary medicine, the author examines the strengths and remaining weaknesses of a correspondingly revamped hybrid account of disease.

Keywords: evolutionary medicine, philosophy of medicine, philosophy of biology, disease, disorder, pathology, function, dysfunction, malfunction

© Peter Takacs 2025

ISBNs: 9781009768061 (HB), 9781009768016 (PB), 9781009768047 (OC)
ISSNs: 2515-1126 (online), 2515-1118 (print)

Contents

1 Walking the Middle Path in the Philosophy of Medicine   1

2 A Menagerie of Attempts to Naturalize Dysfunction   16

3 Medicine *Sans* Evolutionary Biology?   29

4 The Biology of Disorder: A Burgeoning Philosophical Consensus   42

5 From Sensible Normativism to a Hybrid Account   57

   References   75

1 Walking the Middle Path in the Philosophy of Medicine

1.1 Introduction

Medicine is a highly inclusive applied science. It relies on indispensable contributions from various subdisciplines of mathematics, physics, chemistry, biology, and psychology. Within this interdisciplinary admixture, the biological sciences – molecular biology, physiology, anthropology, ecology, and even evolutionary theory – still hold pride of place. Verifying as much requires little more than cursory acknowledgments of the development of the germ theory of disease, the contribution of microbiological and molecular techniques to antibiotic development, and the critical role of biochemical tests (e.g., reagents) during diagnosis, to name but a few examples (Schaffner 1993). The aforementioned areas of biology examine the diverse physiological organization, morphology and behavior of lifeforms, the spatial and temporal distributions of organisms, and how past and present interactions among them and their respective abiotic environments continue to shape prospects for survival. Biology in general aims to explain how and why organisms exist in the myriad ways they do. Medical theorizing, in turn, gleans biological insights about organisms to develop a consensus regarding minimally necessary physical characteristics for human well-being, a set of attributes that must be maintained no matter what particular (reasonable) goals individuals may have. This is the *ideal* toward which modern medicine strives. Diagnosis, prognosis, and intervention presuppose it. Corrective intervention or treatment may well be the general aim of medicine *as a practice*. But a theoretical account of what it is for a human to be "healthy" and, thereupon, what deviations from physiological or morphological normality merit the labels "unhealthy," "disordered," "pathological," or "diseased" serves as a guide for whether and how to intervene.

1.2 The Challenge of Disentangling Pathological Deviation and Extant Variation

This ideal aim of medicine has surprisingly been hampered by the very science of biology itself. Discovery after discovery has revealed that human populations exhibit startling amounts of genetic, morphological, physiological, and behavioral variation. It is increasingly implausible to suppose there is but one optimal way of being human. Any apparent Aristotelian "essences" that might be cited as fundamentally distinguishing humans from other species have proven to be illusory on close inspection. This only becomes more evident when we acknowledge the existence of common ancestry in the distant evolutionary past. An organism is not a member of a species because it has a shared essence with all other members of the species. Rather, the members of a species resemble one another to a greater or lesser extent because they are united by a pattern of ancestry and descent. Seemingly obvious distinctions of kind (species) gradually blur away when we examine the diversity of life as a phenomenon distributed across highly variable populations on phylogenetic (evolutionary) timescales.

Appreciation for the scope of this biological challenge to the aims of medicine requires a brief foray into the history of evolutionary theorizing. A fundamental ontological reorientation occurred when Charles Darwin and Alfred Russell Wallace (Darwin et al. 1858, Darwin 1859/1964]) introduced the theory of evolution via natural selection. Variation in form and function across species was obvious even to their intellectual predecessors. Largely unrecognized until their contributions, however, was the fact that species diversity has its causal precursors in intraspecific variation. Trait variation within distinct populations of a species allows for adaptation to localized selection pressures. Since environmental pressures can vary dramatically among populations, so, too, can the direction and tempo of evolutionary diversification. Ernst Mayr (1961), a renowned evolutionary biologist, captures the implication of this key insight:

> The assumptions of population thinking are diametrically opposed to those of the typologist. The populationist stresses the uniqueness of everything in the organic world. What is true for the human species – that no two individuals are alike – is equally true for all other species of animals and plants [...] All organisms and organic phenomena are composed of unique features and can be described collectively only in statistical terms. Individuals, or any kind of organic entities, form populations of which we can determine the arithmetic mean and the statistics of variation. Averages are merely statistical abstractions, only the individuals of which the populations are composed have reality.[1]

With the acceptance of Darwin's and Wallace's ideas, trait variation was no longer conceived of as unfavorable accidental deviation from a species archetype or essence, as Mayr's "typologist" would assume. It was the very foundation for the bewildering diversity of adaptive form and function in the organismic world.

The ramifications of this ontological about-face for medicine could not be more dramatic. Medical theorizing must accommodate relevant biological discoveries if it is to generate a conception of human health and, as a

---

[1] Op. xix–xx. Facsimile of the first edition of Charles Darwin's *Origin of Species*.

corollary, disease. Evolutionary biology has revealed a bestiary of variation in populations, human or otherwise. Discoveries in physiology only further compound the enormity of the challenge posed by these observations in evolutionary biology. This ubiquitous variation is not incidental; it is the ultimate fodder for adaptive evolution. Successfully accomplishing the overarching project of medicine as a practice (corrective intervention) consequently depends on the ability of its practitioners to generate principled medical reasons for distinguishing between mere extant variation and pathological deviation. Principled medical or scientific reasoning should underwrite the implicit evaluative judgment that accompanies labeling some state "unhealthy."

1.3 A Seemingly Paradoxical Resolution: Uniting Normative and Naturalistic Criteria

While anything but new, it is well worth stressing that the difficulty of establishing a principled distinction between normal variation and aberrance in medicine predates what many readers might take to be the first philosophical incarnations of this problem (e.g., Boorse 1977). Notably, it is prefigured by Georges Canguilhem's criticism of Claude Bernard in *On the Normal and the Pathological* (1943/2012).[2] Bernard maintained that disease is the result of continuous perturbation from an objectively normal and measurable physiological state or range: "Every disease has a corresponding normal function of which it is only the disturbed, exaggerated, diminished or obliterated expression. If we are unable to explain all manifestations of disease today, it is because physiology is not yet sufficiently advanced and there are still many normal functions unknown to us" (p. 68, Canguilhem 1943/2012). Crucially, Bernard's thinking overlooks the possibility that deviations from what is deemed a baseline range of physiological performance or typical physical constitution can sometimes become the "new normal" or equilibrium state for a particular organism.

Sophisticated instances of such "new normality" will be discussed at length in Section 4. For now, the subtle shortcoming of Bernard's "steady state" or "static" view can be illustrated by considering two aspects of human development, namely brain sparing and catch-up growth (Cosmi et al. 2011). The phenomenon of brain sparing occurs during human fetal development when a fetus experiences insufficient amounts

---

[2] It is also worth mentioning that Canguilhem's holistic thinking draws heavily on ideas expressed by the German neurologist Kurt Goldstein.

of oxygen or nutrients in utero. Blood flow in placentally deprived fetuses is preferentially directed toward the brain at the expense of the other vital organs. This makes good sense considering the centrality of cerebral functioning in the nervous system and its importance to the control of human physiology post parturition. Deprived fetuses usually have small body size for their gestational age as a direct consequence of ensuring proper brain development. Being small for gestational age is an important factor in determining whether to deliver preterm via induced labor or caesarian section. One could accordingly regard small size for gestational age as an unambiguously pathological condition. In many instances, it would be fair to characterize it this way since poor nutrition during development can increase disease risk later in life. However, small size for gestational age is not pathological in every case. It can be an adaptive response to adverse but transient nutritional conditions. There are well-known compensatory mechanisms. One of these is "catch-up growth" (high-velocity growth acceleration), which occurs in young children when the causes of a growth deficit are removed and abates when a statistically normal for age growth trajectory is achieved. Identifying a compensatory mechanism like this can reframe the statistically abnormal condition as an adaptively plastic developmental response in certain cases. This type of example suggests a broader and more dynamic domain of normal functioning than Bernard's view seemingly assumes. At least in some instances, the systemic pairing of small size for gestational age (due to brain sparing) with adequate catch-up growth post parturition is another way of realizing a healthy developmental trajectory toward adulthood given the constraint of suboptimal conditions during transient stages of growth.

Canguilhem presciently recognized the possibility of such transformations of normality: "In general, any one act of a normal subject must not be related to an analogous act of a sick person without understanding the sense and value of the pathological act for the possibilities of existence of the modified organism" (1943 [2012], p. 86). Few would nowadays resist the claim that individuals tend to make the best of a seemingly bad situation, whether physiologically and developmentally (unconsciously) or behaviorally (intentionally). Acknowledging the diversity of ways that individuals can adapt to their circumstances must have an important bearing on medical diagnoses. Canguilhem's examination of the implications of biological renormalization for medical diagnosis was among the first to note its importance.

As shrewd as Canguilhem's insight might have been for expanding medical conceptions about the natural range of non-dysfunctional variation, it inadvertently raised another difficulty for principled determinations of disease or medical disorder. A distinctly evaluative sense of normality, as involving moral norms or values, is conspicuous in Canguilhem's stress on "understanding the *sense and value* of the pathological act for the possibilities of existence *of the modified organism*" (italics added). This insinuates that medicine must also account for an individual's assessment of well-being before it deems a condition pathological. The challenge accordingly becomes one of determining whose "sense(s) and value(s)" deserve precedence when judging whether a human organism "makes the best of an otherwise bad situation." The evaluative basis for this judgment might, for instance, reside exclusively with an individual's subjective self-assessment of her/his condition. Individuals would then be the final arbiters for all matters pertaining to whether they do well enough in adverse circumstances and, therefore, whether they are (un)healthy. The notion of *illness* would consequently feature centrally in determinations of *disease*. Many in the medical community might recoil at this possibility since pathologies are typically defined on the basis of external (objectively verifiable) criteria, whereas illnesses are more commonly identified by reference to the (self-reported) internal experiences of individuals. Alternatively, decisions concerning whether one has made the best of an otherwise bad situation might rest primarily with some sort of medically unprincipled, intersubjective consensus (e.g., a health fad). This type of relativism about disease determination is no less concerning in the eyes of the medical community. As the rampant spread of misinformation during the recent COVID-19 pandemic has shown, norms adopted by collectives (e.g., communities) can prove even more difficult to eradicate than those held by one or a few highly idiosyncratic individuals. Most medical practitioners and many philosophers of biomedicine share the belief that neither Stoic subjectivism nor any mere tyranny of convention should figure among the medical reasons for determinations of pathology. Canguilhem's proposals bring into sharp relief the difficulty of cogently articulating a set of principled reasons for overruling the "senses and values" of individuals, communities, or even societies when their health norms do not directly align with those established by the medical community.

    A word of warning to readers. In this Element, I will countenance the internal experiences of individuals (i.e., self-reported illnesses or individual assessments of well-being) only insofar as these experiences aggregate non-randomly in a communal or societal setting. The instrumental value that an individual associates with his/her physical condition can obviously depend on many factors. It depends importantly on what "possibilities of existence" (to borrow Canguilhem's wording) remain open and the rank ordering that an individual imposes on these available forms of existence. It depends no less crucially on the extent to which an individual sees their condition as promoting or inhibiting the realization of their preferred ranking. Be all of this as it may, the only genuine way *to convincing others* that one has in fact "made the best of an apparently bad situation" minimally requires comparison of this professed success against the assessment of others who share both the same (or a directly comparable) condition and a nearly identical rank ordering of preferences. In short, assessment must be broadly intersubjective in character if it is to have any significant bearing on social deliberations surrounding health. In this limited sense, I follow directly in Canguilhem's footsteps. For all his criticism of those who seek a completely mind-independent or convention-independent conception of disease, Canguilhem was no run amok subjectivist. He understood full well that physicians and medical insurers, unlike many who work in the service industry, must routinely defy the slogan "The customer is always right." The experiential distress of an individual can often be untethered from a patient's knowledge about the causal sources of that distress. Where I depart from Canguilhem's thinking will emerge later.

1.4 Transitioning into the Contemporary Debate: Naturalism versus Normativism

Many philosophers of medicine have drawn the conclusion that scientifically established health norms in the medical community cannot be reconciled with social health norms whose roots lie elsewhere. We shall examine some of the reasons for this (undue) pessimism and its implications later in this section. Let us, for the time being, focus instead on the possibility of reconciliation. How might reconciliation take shape? At least since Canguilhem's work, it begins by acknowledging that there are two distinct dimensions of normality associated with ascriptions of disease. Clearly identifying each dimension sheds considerable light on how they might relate to one another.

The first dimension of normality concerns how judgments about pathology must account for circumstances or features that are unique to subgroups of individuals. Attributions of health and disease often depend crucially on how a population is statistically partitioned and whether medically relevant partitions (citing recognized causative agents) can change the way we view the prevailing range of functional performance. High blood pressure (hypertension) furnishes us with a good example. The current medical threshold for hypertension is any measure that meets or exceeds a resting systolic pressure of 130 (mm Hg) or a diastolic pressure greater than or equal to 80 (mm Hg) (Whelton et al. 2018). For brevity's sake, these measurements are commonly expressed as a ratio of systolic to diastolic measures, high blood pressure being correspondingly expressed as 130/80 or "130 over 80."[3] It is well established that untreated hypertension can cause cardiovascular complications such as heart disease and stroke. But hypertension is not uncommon for women during pregnancy. Some form of hypertension occurs in 5–10% of pregnancies worldwide (Regitz-Zagrosek et al. 2018). Among those who experience pregnancy-induced hypertension, many exhibit a form called "gestational hypertension." These women have high resting blood pressure (≥140/90) that develops only after twenty weeks of pregnancy without excess protein in the urine or other signs of organ damage as with more severe forms of hypertension (e.g., preeclampsia). Gestational hypertension can be a risk factor for heart disease (Riise et al. 2018). However, in the absence of other complications and comorbidities, additional monitoring and relatively minor lifestyle changes to diet or physical activity are usually the medically recommended courses of action.

Following Canguilhem's line of thought, resting blood pressure that exceeds 130/80 could be construed as just another way of achieving "normal" or healthy regulation given the physiological circumstances of pregnant women without additional complications. Other findings, such as those of the Framingham Heart Study (a longitudinal study now spanning three-quarters of a century), further complicate any attempt to identify an unchanging "normal" value or range for resting blood pressure by revealing that readings even below the current upper limit set by health care workers might be deemed a risk factor for cardiovascular disease if one is, for instance, morbidly obese and a heavy smoker. These examples, which could be easily multiplied, show how the range of non-dysfunctional variation in performance can be surprisingly large, malleable, and crucially dependent on other relevant facts about groups of individuals or their circumstances. Easily implementable "static" guidelines for identifying the pathological, while perhaps practical from a clinical standpoint, can obscure this range of natural variation. When performance conditional on variation in other relevant (potentially synergistic) causal factors goes

---

[3] While few would assign significance to *the ratio* of systolic to diastolic pressures, I maintain conventional notation in order to prevent misunderstandings.

unacknowledged, medicine runs the risk of assigning pathology either far too readily or not often enough.[4]

The second dimension of normality concerns how judgments of disease presume some preexisting standard of propriety. The worry along this dimension is that medical judgments pertaining to standards for health may not be quite as independent of social norms as is sometimes supposed. Medical characterizations of homosexuality are routinely deployed to legitimize this concern. Homosexual orientation or behavior has historically been viewed as aberrant if not an outright pathology (DSM-I 1952; ICD-9 [Slee 1978]). It is debatable whether the term "homosexuality," even as used today, designates a single trait or any determinate suite of behaviors. The collections of phenomena (genetic, behavioral, or anatomical) that have at one time or another been associated with homosexuality may just be unprincipled conglomerations of heterogenous features. Although an important consideration, especially but not exclusively in the domain of psychiatry, nothing hangs on this issue of reification here. We can proceed *as if* homosexuality is in fact some objective biological phenomenon in order to elucidate worries about the incursion of social norms into medicine. It was initially classified as a "paraphilia" (DSM-1 1952), a disorder in which one's sexual arousal and gratification depend on fantasizing about and engaging in sexual behavior that is atypical and extreme. Although later editions of DSM-II (eighth edition) showed a burgeoning shift in its classification from a paraphilia to a "sexual orientation disturbance," there was no substantive change in its classification until 1973 when the Australian and New Zealand College of Psychiatry Federal Council and the American Psychiatric Association deemed that homosexuality was not a mental illness. Early evolutionary views and misconceptions about evolutionary theory were also occasionally complicit in portraying homosexuality as a disorder (Ruse 1988; Murphy 1997). It was, for example, long held that homosexuality could not be anything other than maladaptive because it tends to reduce the frequency of heterosexual encounters and thus lower the reproductive output of individuals who engage in such behavior. Largely in agreement with then-prevailing cultural and psychiatric norms, homosexuality was theoretically maligned as a self-eradicating reproductive strategy. Even this truncated history of homosexuality shows just how

---

[4] Similar worries about whether a condition is in fact pathological for an individual and how to tailor treatment specifically to individuals now motivate the burgeoning Precision (Personalized) Medicine movement. It receives scant attention in this Element for the simple reason that it is too early to tell whether the movement can deliver on its promises.

tempting it is to maintain that the opinions of individuals or small minorities should feature fundamentally when it comes to disease determination. For those who sympathize with this line of thought, the subjective, holistic perspective of the individual whose disease status is in question supposedly has medical value insofar as it serves as a bulwark against potentially misguided determinations of the medical establishment.

Two distinct senses of normality can influence medical decisions about whether a condition is diseased. One deals with the epistemic issue of which causative agents and environmental factors to consider when establishing ranges for healthy functioning. The other concerns the incursion of non-epistemic social values into medical theorizing and practice. Both impose antecedent qualifications upon medicine's *ideal* aim of establishing hard boundaries on a conception of health. These push the practice of medicine ever farther from a categorical conception of health and pathology. Whereas, prior to Canguilhem's work, one might have made the case for blanket definitions of any pathology in roughly the form "A patient exhibits pathology X if and only if an intrinsic physical process or attribute brings about an effect beyond some determinate duration or range," the criteria for most subsequent diagnoses must include sets of clauses regarding the acceptable range(s) of performance for specific physical attributes across a variety of environmental contexts and perhaps even social circumstances for distinct subgroups.

Based on striking examples such as the medicalization of homosexuality, many have subsequently seen fit to echo the sentiments of Rudolf Virchow, founder of cellular pathology, who (in)famously averred that "Medicine is a social science and politics is nothing else but medicine on a large scale" (Taylor and Rieger 1985 [1848]). Virchow's provocative quote endorses a very strong deflationary position when it comes to the two dimensions of normality involved in judgments of pathology. It controversially implies that there cannot be a purely descriptive, objective scientific basis on which to establish a principled distinction between health and disease. For we supposedly cannot rule out the possibility that science unwittingly reflects norms into nature and later rediscovers those norms in the guise of "objective" truths that then serve to justify existing social norms (Ruse 1996). In effect, the dimensions of normality that Canguilhem's work helped us distinguish would then collapse into one. What allegedly remains is an expansive range of natural biological variation over which a spectrum of social norms and political power dynamics prevail, the medical and scientific community at large purportedly being no exception.

Two common commitments define this popular relativistic position today. First, it is committed to the idea that facts about health or pathology can only be determined via normative considerations. This renders it strongly normativist in the sense that human values, beliefs about what goals and means of achievement are good (just) or bad (unjust), ultimately ground evaluations of biological variation as pathological or healthy. Normativity is supposed to be a matter of kind rather than degree from this perspective. If facts about health and pathology are even minimally normative, then they must be wholly and fundamentally normative. Second, the ontological status of facts about health or pathology are necessarily mind-dependent (subjective) or convention-dependent (intersubjective). Health facts are essentially human constructs, and the constructing relation is supposed to be constitutive. It is part of the very nature of such constituted objects that they depend substantively on humans. There can be no pathological physical states without some form of human judgment to the effect that such states are disordered. Taken together, these two commitments carve out a broad position that I will refer to as "Subjectivist-Normativist-Constructivist," or "SNC" for short (Margolis 1976; Goosens 1980; Sedgwick 1982; Engelhardt 1976; Brown 1990; Foucault 1973; Cooper 2002; Glackin 2010). Readers should note that any unqualified mention of Normativism hereafter is intended to be synonymous with SNC. I shall later (Section 5) distinguish two currently influential varieties of SNC, which I dub "Internalist" and "Externalist." For the moment, the distinguishing features of these two forms of Normativism can be ignored.

Not everyone is comfortable with the idea that judgments concerning our biology must be so strongly normative. Unearthing potential reasons for disagreement is not difficult.

One can immediately question the desirability of giving due consideration to the goals of individuals or subgroups prior to diagnosis. Given idiosyncratic physiological or environmental circumstances, not to mention the bewildering diversity surrounding conceptions of the good life, it may in the end prove impossible for Normativists to avoid a radically subjectivist account of health. Some Normativists seem to imply that unfettered liberty of health determination is not a serious concern on the grounds that a sensible social majority would quash it in most cases. But, as these Normativists must surely realize, this is not always so. We need look no further than the medicalization of homosexuality to show that there are sometimes no principled means by which to limit the power of unjust social or institutional arrangements. It seems that Normativists must find a way to limit radically (inter)subjective conceptions of health and disease. But if majority rule is their proposed preventive for radical (inter)subjectivism, why are the disease determinations of a social majority not given precedence over minority opinion in every case, including what most would now see as the wrongful characterization of homosexuality as a medical disorder? For Normativists, there is apparently no recourse to a separate domain of "pristine" (objective) medical reasons. Any such domain would, by their own lights, be riddled with non-epistemic social values. Dominic Murphy (2015) has captured the crux of the problem with admirable clarity:

> The chief problem for Normativism is that we routinely make a distinction between the sick and the deviant, or between pathological conditions and those that we just disapprove of. Our disease concepts cannot just be a matter of disvaluing certain people or their properties. It must involve a reason for disvaluing them in a medical way rather than some other way.

There must be a principled way to prevent a "tyranny of the masses" as well as radical subjectivism when it comes to disease determination.

Another difficulty faced by Normativist accounts of health is that biologists routinely distinguish between normal and pathological variations when it comes to nonhuman organisms. Their distinctions are not formed on the basis of intuitions that project human values or social norms onto their (nonhuman) subject matter. Rather, biologists understand that neglecting the difference between healthy and pathological specimens threatens to undermine the explanatory and predictive integrity of their science. John Matthewson and Paul Griffiths (2017) have made this point more forcefully than most:

> When a biologist encounters a new species of beetle, a core part of his or her work will be determining which specimens are relevant to describing the biology of that species and which are only relevant to the pathologist. For example, beetles squashed during collection, or half-eaten by a predator, should not be included in the description of the new species. Those which differ from their conspecifics because they have a high parasite load need to be distinguished from the healthy specimens, not lumped in to determine an average value for some phenotype (p. 451).

In light of such considerations, those who enthusiastically endorse Virchow's analogizing medicine to politics would do well to revisit another important strand of his thinking: "Between animal and human medicine, there is no dividing line – nor should there be" (1858).[5]

---

[5] Virchow, among others (William Osler), are the historical precursors for what has come to be known as the "One Health" or "One Medicine" movement. Interested readers should consult Rabinowitz et al. (2017).

Based on these difficulties, a number of thinkers have seen fit to conclude that Normativism should be rejected wholesale (Boorse 1977; Wachbroit 1994; Kitcher 1997; Thagard 1999/2018). In its place they offer what I will call an "Objectivist-Naturalist-Realist" alternative ("ONR" for short). Unless otherwise noted, the term "Naturalism" will be used as a stand-in for this broad view (ONR). The distinguishing ideological commitments of this competing account of health and pathology are articulated in explicit contrast with those of SNC. Facts about human health and disease are proposed as purely objective phenomena, having a mind- or even convention-independent reality. These facts are subject to discovery as opposed to being mere constructions. According to Philip Kitcher, "objectivists about disease, think that there are facts about the human body on which the notion of disease is founded, and that those with a clear grasp of those facts would have no trouble drawing lines, even in challenging cases" (1997, pp. 208–209). Christopher Boorse, who has arguably been the quintessential exponent of Naturalism, states this commitment in the strongest possible manner: "On our view, disease judgments are value neutral [...] their recognition is a matter of natural science, not evaluative decision" (1977, p. 543). As Section 2 will make clear, his basic contention is that health and disease can be determined in a purely quantitative fashion by reference to statistical thresholds for variance and deviations exceeding those boundaries. Only the more limited *epistemic* norms of science, or what are now commonly referred to as "cognitive or theoretical virtues/values," would then be relevant for assessing medical determinations (Kuhn 1977/2011; McMullin 1982).

1.5 A Fruitless Dichotomy: Motivation for a Hybrid Account of Disease

The philosophy of medicine has been dominated by a dialectic in which versions of SNC (Normativism) are pitted against variants of ONR (Naturalism). This conceptual trajectory has culminated in an oversimplified and potentially harmful stalemate between these two allegedly exclusive and exhaustive positions. Both positions are needlessly extreme and suffer conceptual drawbacks. I will conclude this section by presenting two philosophical arguments for dissolving the prevailing dialectic.

The first argument takes aim at the problematic assumption that (non-epistemic) normativity is an "all or nothing" affair. In different ways, both Normativism and Naturalism depend on the fact that it is. Normativists generalize from the fact that the biomedical sciences have on occasion adopted prevailing social norms to guide identification of the pathological to the conclusion that it is very often if not always so, even if medical practitioners are unaware of it. Naturalists attempt to counter this sweeping extrapolative inference by demonstrating that social norms need not feature among the criteria for determining human or nonhuman pathologies. Either way, if judgments regarding health and pathology are in any way value-laden, it supposedly follows that health facts must be mere constructs that lack any objective (mind-independent) reality. Facts about health are allegedly either normative or naturalistic, but certainly not both.

This false dichotomy continues to fuel the argumentative dialectic in the philosophy of medicine. It is false because there are two *orthogonal* dimensions at work (Broadbent 2019). The first involves metaphysical considerations of objectivity and epistemological concerns about how we might come to know about the world. It addresses whether facts about health are extrinsic to the minds, cultures, or societies that identify such ailments. This dimension overlaps roughly with the issues raised by the epistemic sense of normality alluded to in Canguilhem's work. The second dimension concerns the extent to which judgments of medical disorder are dependent upon non-epistemic norms and values, which would include the "senses and values" of modified organisms as noted by Canguilhem. These are *logically independent* considerations. That facts about health are value-laden does not entail a lack of objectivity. Some moral realists, for example, hold that moral facts are independent of what anyone may think about them, and that these facts are not discoverable via any sophisticated and complete form of empirical inquiry (Shafer-Landau 2003). Moral facts, on this view, are thus objective but nonnatural. Realists of this ilk can accordingly accept that a notion of disease is so deeply intertwined with our ideas of what is good or bad for a person that it simply cannot be a matter of scientific fact alone, while nevertheless denying that this amounts to mere constructivism or conventionalism (Stempsey 2006). But if value considerations do not entail nonobjectivity, then (by contraposition) objectivity does not entail the absence of value considerations. Facts about health and pathology can apparently be both normative and objective, contrary to the commitment that has figured so centrally in arguments for both Normativism and Naturalism.

This logical rejoinder to the prevailing dialectic in philosophy of medicine can be bolstered by a second argument that emphasizes more comprehensive historical investigation into judgments of human health and pathology. The case of homosexuality yet again proves exemplary. Recent evolutionary theorizing has proposed mechanisms that can offset the apparent selective costs of homosexuality. Biologically meaningful interpretations of W.D. Hamilton's model of kin selection as well as Robert Trivers' model for reciprocal altruism offer (competing) explanations of the relatively high cross-cultural prevalence of homosexuality. With respect to the former, it has been argued that the fertility cost of male homosexual behavior is offset by a fecundity advantage to the mothers of homosexual or bisexual children (Camperio-Ciani et al. 2004; Iemmola and Campiero-Ciani 2009). In the latter vein, Roughgarden (2012a, 2012b) contends that homosexual behavior may be a special case of physically intimate behavior that promotes cooperation through the reciprocal sharing of pleasure. What was once believed to be a disorder by layperson and scientist alike has been reconceived (by some) as a potentially adaptive behavioral strategy. The adaptiveness of homosexuality remains a hotly contested theoretical issue. Depictions of it as a selectively maintained strategy are still viewed with suspicion by many in the evolutionary community.[6] But whether or not it turns out to be adaptive is immaterial for my purposes. What matters is only that we now have principled theoretical reasons for seriously entertaining the possibility.

What does this diachronic extension to the history of homosexuality as typically portrayed by Normativists reveal about the metaphysical objectivity of health facts? Despite considerable entanglements with prevailing customs and norms in many societies, views that regard homosexuality as aberrant or unhealthy have been increasingly strained in light of more recent evolutionary theorizing and the discovery that homosexuality is far more common than once believed. These scientific advances (novel theoretical hypotheses and empirical discoveries) were not, of course, the catalysts for social change. They followed on the heels of broad social movements in some societies. It would nevertheless be incorrect to conclude that these recent scientific proposals simply reflect the values of a newly emerging status quo with respect to sexual orientation in such societies. For these recent proposals cut just as deeply against the hope of some gay-rights advocates for the discovery of a "gay gene" (i.e., allelic variants of large effect) as against those who would depict homosexual orientation as a learned behavior and, perhaps, poor "lifestyle choice." This example shows that, at least in some societies, science appears to have considerable independence from sociocultural factors as well as resistance to unadulterated constructivism. Epistemic norms and the aims of scientific inquiry can hold moral values and social norms at bay.

---

[6] I thank an anonymous reviewer at Cambridge University Press for pressing this point.

1.6 A Philosophy of Biomedicine for the Twenty-First Century

The oft-overlooked absence of logical entailment and a reliance on artificially delimited (synchronic) historical episodes are disconcerting. No less disappointing is that both Normativists and Naturalists often availed themselves of what is but a partial depiction of the biological sciences informing modern medical praxis. The details of this allegation form much of the content in the sections to come. Suffice it to say, at present, that caricatures of biological science, especially when it comes to portraying evolutionary biology, have too often licensed the premature dismissal of attempts to naturalize the notions of health and pathology.

This Element is a short but sustained attempt to redress this unfortunate situation in a way that does not fall prey to the conceptual tangles that have undone ONR (Naturalism) in the past. More than anything, I would like to convince readers that we should abandon the prevailing dialectic in the philosophy of medicine and instead focus on the development of a weakly normative hybrid position. The hybrid "Life History-Organizational Account" (LH-OA) of disorder I propose (Section 4) is not without precedent. Hybrid accounts of disease have become respectable since Jerome Wakefield (1992a) introduced his "harmful dysfunction" account of psychiatric disorders. Others have also taken up this hybrid mantle, albeit to draw strikingly different conclusions (Stegenga 2018).[7] Even the form of Normativism that I call "Internalist SNC" (Section 5) might be considered "hybrid" in its own limited way. What such distinct views agree on is that the most promising way forward in the philosophy of medicine demands some degree of hybridity. In most general terms, an account of health qualifies as hybrid only if the identification of disease or disorder requires both an objective scientific component and a (inter) subjective value component. Each component is necessary; jointly they are sufficient. The hybrid account developed in this Element takes a serious but measured approach to the implications of recent biological insights for medical theory and practice. However, it never forgets that values of some sort are ineliminable.

Herein lie the outlines of a more scientifically and philosophically nuanced version of the harmful dysfunction account. There is conceptual

---

[7] As I understand Stegenga's so-called "Master Argument," he favors a hybrid method whereupon either curing the disease or caring for the harm the disease causes is sufficient to establish the medical intervention's effectiveness. However, he ultimately argues for a form of "medical nihilism" because he believes that most medical interventions fail to do either. I find this pessimism questionable.

room for such a position. While the account on offer undoubtedly confronts its own set of challenges, it is well worth articulating what these challenges are and the prospects for meeting them.

2 A Menagerie of Attempts to Naturalize Dysfunction

2.1 The Gradual Turn Toward an Evolutionary Account of (Dys)Function

Section 1 laid bare a fundamental conceptual challenge for medicine: what principled reasons enable us to identify physical variations as pathological? This vexing issue cannot be shirked off as merely of philosophical concern; it arises as a direct consequence of the science – biology – that most directly informs modern medicine. Hybrid accounts (Wakefield 1992a, 1992b) of the kind I defend in this Element contend that objective criteria for health and disease exist independently and harmoniously alongside of normative considerations. The cogency of any hybrid account thus partially hinges on whether there are any scientific criteria for human disease/disorder that can resist the influence of broader (non-epistemic) social norms. Many have turned their attention to examining the theory and methods of the biological sciences in order to determine whether any scientifically established norms of medicine have this sort of fortitude (Nesse 2019; Neander 2017; Garson 2019; Matthewson and Griffiths 2017; Shea 2013).

The predominant reason for this turn is that biology already includes a principled conceptual distinction that parallels and potentially justifies the much sought-after distinction between health and disease. The pivotal distinction is that between function and dysfunction/malfunction. Evolutionary biologists routinely distinguish between functional and dysfunctional traits when determining species membership (Matthewson and Griffiths 2017). Doing so requires them to conclude that some organismal characteristics are incidental to class membership and can thereby be considered abnormal. In similar fashion, physiologists regularly distinguish biophysical and biochemical systems or processes that function normally from those that compromise the integrity of organisms. Whenever there is informed discussion of human medical disorders or diseases, it is taken for granted that there must be a root cause of physical distress or suffering. The basic intuition is that "something is broken inside" (the physical boundaries of) an individual. A constituent part or intra-organismic process somehow fails to bring about the effects that maintain an organism's physical well-being. This is just another way of suggesting that there has been some sort of malfunction. Implied dysfunction presupposes an account of normal functioning. Evolutionary biology and physiology, with their respective emphases on natural selection as a causal process that maintains traits that tend to increase organismal fitness and the homeostatic regulation of organismal integrity, stand out as a particularly promising areas for grounding intuitions about functionality.

The general aim of this section is to sketch a basic conceptual topography of attempts to ground normal functioning and consequently dysfunction in a way that does not immediately invoke broad (nonmedical) social norms or values. I will scrutinize these alternative philosophical accounts of biological function along the way. The dialectic pitting these Naturalistic (ONR) accounts of function against one another merits emphasis. Their perceived failure, as a collective sharing a common goal, has prompted many thinkers to conclude that some variety of Normativism (SNC) offers a better account of health and disease.

2.2 A Hitchhiker's Guide to Naturalizing (Dys)Function: Precursors to the Evolutionary View

The Causal Role account of function (Cummins 1975; Amundson and Lauder 1994) offers us the most direct philosophical articulation of the intuition that pathology involves something being broken inside an individual. The target of analysis on this account is a capacity of the system or subsystem to which a component belongs. A systemic capacity is to be explained in terms of the causal contributions that its components make to the exercise of the capacity. The human heart, seemingly as always, provides an illustrative example. It causally contributes to the capacity of our circulatory (cardiovascular) system for removing waste products from tissues and delivering nutrients and oxygen. How does the circulatory system manage to exercise this capacity? To a large extent it is accomplished via the heart's functional (causal) role as a pump within the system. The heart's capacity to pump blood can, in turn, be decomposed into contributions of subcomponents (e.g., chambers and valves) that then explain how it is that the heart pumps. This process of identifying and descriptively decomposing nested capacities can be repeated almost indefinitely. Below the biomolecular level, however, one would presumably no longer be explaining an unequivocally biological phenomenon.

The Causal Role account presents us with an admirably straightforward pattern for answering questions about the function of a component: The function of a part or process is to bring about whatever effects it happens to cause in any of the (sub)systems partially comprised by it. This strategy leaves considerable leeway for researchers when it comes to "localizing" (i.e., "identifying a locus of control" as per William Bechtel and Richardson (1993)) systems or capacities that are of explanatory interest. A component can make causal contributions to many distinct systems and thus consistently sustain a multitude of function ascriptions. Note, too, that there is no explicit mention of the historically distant, more recent, or even current evolutionary purpose of a biological component. The Causal Role account does not require appeal to evolutionary considerations. Once the capacity of a (sub)system is identified as being of explanatory interest, analysis proceeds in purely descriptive, causal terms.

There is nevertheless a significant drawback to the Casual Role account that casts doubt on its characterization of explanatory strategies in biomedical research. While it may provide an account of a component's function via describing its *actual* causal contribution(s) to the capacity of some system, it is initially silent about whether a component is performing the way that it *should*. Advocates for the Causal Role account are typically unfazed by this absence of patently teleological and purposive language. For them, the prescriptive nature of purposive (function) claims has only a heuristic role (Schaffner 1993, p. 410). But without a convincing story about why a systemic capacity has the particular range of "normal" or "appropriate" activity that it does, the Causal Role account cannot justify the assertion that a component is dysfunctional. It is unable to distinguish the causal contributions that a component *happens to make* from causal contributions *required for* (sub)systemic success.

The Causal Role account aims to meet this challenge by stipulating the functional norms of the system being analyzed. That systemic norms are conditionally stipulated at the very outset of a Causal Role analysis does not mean that such norms lack justification altogether. Background knowledge of collateral theories that refer to a system and its structure can often be invoked to support their assumptions. Consider, for example, built-in presuppositions about the human circulatory system's capacity to deliver nutrients and oxygen as well as remove waste products from its tissues. It is taken for granted that this capacity can be exercised effectively only over a limited range. No one would dispute the fact that the heart makes a significant causal contribution to this circulatory capacity. But the heart is not functional merely in virtue of the fact that it happens to pump; it must also pump *adequately*. To pump adequately, however, is to pump in a way that delivers nutrients and oxygen and removes waste products over a specified range, namely one that does not compromise organismal health. The attribution of function to the heart consequently hinges on *physiological norms* regarding the appropriate exercise of a capacity.

Even though justified on theoretical and empirical grounds, the resort to physiological norms pushes questions about proper functioning back a step, from the component part to its embedding system(s). The norms of functioning that a system imposes on components must also be justified by appeal to yet more inclusive systems that contain the initial system as a component (subsystem). A regress looms. When it comes to medicine, the overarching systemic norms that govern function ascription for all parts or processes ultimately belong to the organismal type (*Homo sapiens*) whose health is in question. This point has been made before: "The proper functions of a biological trait are the functions it is ascribed in a [Causal Role] functional analysis of the capacity to survive and reproduce (fitness) which has been displayed by animals with that feature" (Griffiths 1993, p. 412). Even the author of the Causal Role account, Richard Cummins, noted as much: "A more plausible suggestion along these lines in the special context of evolutionary biology is this: (8) The functions of a part or process in an organism are to be identified with those of its effects contributing to activities or conditions of the organism which sustain or increase the organism's capacity to contribute to survival of the species" (1975, p. 755). Setting aside its misplaced focus on the capacity to contribute to *species* fitness rather than *individual* fitness, this quotation reveals that Cummins considered function ascription in biology and medicine to be special instances ("contexts") of the Causal Role account. In a limited sense, this is not wrong. Evolutionary and even physiological functions can be consistently depicted as proper subsets of Causal Role functions. But the pressing question is whether any extension of function analysis in medicine would require going beyond the ultimate systemic norms (survival and reproduction) derived from evolutionary theorizing. If the Causal Role account can find ultimate justification only in the most general systemic goals for biological entities (survival or reproduction), then it fares no better than an evolutionary account of the kind that many believe it supposedly subsumes when it comes to providing a naturalistic account of dysfunction.

In spite of these shortcomings, the Causal Role account can be elaborated in a way that makes it more directly applicable to medicine. It dovetails with a highly influential naturalistic account of health and disease known as the "Bio-Statistical Theory" (Boorse 1977, 1997, 2014). Unlike the Causal Role account, the Bio-Statistical Theory is not after a completely general analysis of the concept of function along mechanistic lines. Rather, its sole aim is to analyze the *theoretical* concept of disease in the sense of "pathological condition" (Boorse 2014, p. 688). The modifier "theoretical" shows that Boorse's intention is to *explicate* (in the sense of Carnap 1950) the concepts of health and disease according to the way that these notions are systematically deployed in the biological sciences, particularly medicine. His analysis is not designed to capture the colloquial usage of these concepts.

The general strategy of Bio-Statistical Theory links the normal functioning of biological components and the health of organisms to statistical normalcy. What is statistically typical always depends on the reference class in question. The explanatorily relevant references classes are supposedly "natural classes" consisting of organisms of uniform functional design; specifically, an age group of a sex of a species (Boorse 1977, p. 555). The normal function of a part or process for members of a reference class is then a statistically typical contribution by it to individual survival and reproduction. A disease or pathological condition is then an internal state that is either an impairment of normal functional ability below typical efficiency for a particular segment or cohort of a population, or a limitation on functional ability caused by a statistically atypical environment. The notion of health consequently becomes a direct corollary of disease: health is nothing more than the absence of disease or pathological condition (Boorse 2014, p. 684).

Parallels with the Causal Role account should be apparent. Identifying mutually exclusive segments or cohorts (Boorse's "natural classes") establishes new systemic types indexed by time (age) and sex. Only within these more precise systems can the exercise of a species-specific capacity for survival or reproduction be fruitfully decomposed. The causal contribution made by a token instance of a component type (e.g., testosterone level) shared by members of a species (*Homo sapiens*) can be fitness-enhancing and thus functional by the standard for one reference class (males aged 15–35) but dysfunctional by the standard in another (females aged 35–55). Were normal functions only those that are statistically typical for the entire species, the Bio-Statistical Theory could not account for group-specific variations in healthy function.

While the Bio-Statistical Theory presents an impressive explication of health and disease, it still suffers from what some see as insurmountable difficulties (Amundson 2000; Cooper 2002; Ereshefsky 2009; Kingma 2010). Perhaps the most potent criticism points out that what is functional (fitness-enhancing) within a natural class can nevertheless be statistically atypical in specific situations. Sickle-cell anemia affords us a good example (Flint et al. 1998). Sickle-cell anemia is a recessive genetic disorder that is often fatal in childhood. It requires being homozygous recessive for the alleles that determine faulty (sickle-shaped) hemoglobin-S rather than normal hemoglobin-A. Those with sickle-cell *trait*, in contrast, are heterozygous at the locus for this phenotype but not anemic. They are merely "carriers." In malaria-ridden regions, those with the sickle-cell trait have a distinct advantage over both homozygous dominant (normal) and homozygous recessive (anemic) conspecifics.[8] Unlike their normal hemoglobin counterparts, carriers of the sickle-cell trait are resistant to malaria because the parasites that cause this disease are killed inside sickle-shaped (S-hemoglobin) blood cells, of which they have some. But this selective advantage cannot push the sickle-cell trait to fixation in the population. Heterozygotes cannot breed true to genotypic or phenotypic form. Despite the increased probability of mating with other carriers in malaria-ridden regions, heterozygotes will produce homozygous (dominant and recessive) offspring in predictable proportions.

How might the Bio-Statistical Theory handle this situation? On a global scale, the sickle-cell *trait* and the recessive sickle-cell allele are at a selective disadvantage and, thereby, statistically less common. Most regions have a negligible malaria burden. It should, therefore, be classified as a dysfunction according to this theory. In some malaria-ridden regions, however, there can be a selective advantage that maintains the trait (heterozygote carrier) at a higher relative frequency than competing (homozygous) phenotypes. This selective advantage can hold even after partitioning populations by sex and age into what are supposedly "natural classes." For any local community with a high burden of malaria, then, the sickle-cell trait can be statistically common and so not a disease because it functions in an adaptive way. Adopting local populations with a high malaria burden as the proper frame of reference for assessing fitness and functionality consequently appears to be the better alternative for resolving the inconsistency faced by the Bio-Statistical Theory in this case. We would not, after all, wish to label the sickle-cell *trait* as dysfunctional in a community where its relatively high prevalence is due to a selective advantage. But the initial problem then reemerges in a modified form for the Bio-Statistical Theory. For it is having just a single copy of the

---

[8] Readers should note that the hemoglobin Hb^A (normal hemoglobin) and Hb^S (abnormal hemoglobin) alleles are usually described as "codominant," rather than as "dominant" and "recessive," because both alleles are expressed and give rise to functional protein products. I adopt the terms "dominant" and "recessive" to ease expression.

recessive sickle-cell allele that confers a selective advantage to carriers. The recessive allele has lower relative frequency than the dominant allelic variant because it is predominantly found in heterozygote form and only rarely if at all in sexually mature individuals with the genotype for anemia. In other words, the statistically uncommon allele is what confers selective advantage in communities with a high malaria burden. Should it then bear the labels "dysfunctional" and "pathological" as the going view would imply? It seriously strains credulity to maintain that it does.[9]

As if the foregoing objection was not troublesome enough, there is yet another fundamental difficulty for the Bio-Statistical Theory. The account seemingly falls afoul of an important condition known as "dispositional adequacy" (Kingma 2010). The performance of a function must be defined in dispositional terms as a capacity or ability because there are many biological traits and processes (e.g., aspects of the immune response) that, at any given moment, are not actively performing the functions that causally contribute to maintaining or enhancing organismal fitness (Kingma 2010, pp. 6–7). Proponents of the Bio-Statistical Theory presumably do not want the nonperformance of such latent abilities at any point in time to automatically qualify as dysfunctional. Their difficulty is that one and the same measure (statistical normalcy) must be applied to determine the normal species function of a component within a natural class not only as it is performed in a specific situation but also for determining how the component would normally be disposed to function in nonactual but biologically realistic situations. A concrete example aids comprehension here. Consider carbon monoxide poisoning. A high level of carbon monoxide usually makes for an uncommon but harmful situation. In the event of carbon monoxide poisoning, hemoglobin has a much-reduced capacity to bind oxygen. This reduction in efficiency is statistically typical given exposure to high levels of carbon monoxide. Hemoglobin's functional state of much-reduced efficiency is then (statistically) normal in this harmful situation. The functional inefficiency of hemoglobin is also statistically typical from the dispositional standpoint. Nearly every individual would exhibit functionally inefficient hemoglobin if exposed to high levels of carbon dioxide. Functional inefficiency is thus statistically typical on either of the ways – actual or dispositional – that are open to judging it pathological. The overall state of the poisoned individual must correspondingly

---

[9] Boorse has addressed sickle-cell anemia. He does not, however, explicitly consider the additional complication of negative frequency-dependent selection. He ultimately concludes as follows: "Given the range of possible heterozygote examples, I suspect the truth is that in heterosis, the medical concept of disease is beginning to show signs of strain" (1997, p. 90).

be considered "healthy" according to the Bio-Statistical Theory (Kingma 2010, p. 11). This clearly runs counter to general intuition as well as informed medical opinion.

Although the Bio-Statistical Theory may still have its defenders, the case against it is daunting. Its demise has been very long in the making, which is a testimony to the sophistication that it introduced into early debate. It has arguably been the most dominant naturalistic theory of health and disease in the philosophical literature. Such has been its influence that even now, more than four decades after its inception, many assume that it is the best if not only hope for naturalizing disease or disorder. There is no shortage of examples expressing this sentiment in the philosophy of medicine. Here is but one provocative aperitif: "In the eyes of many, the refutation of the biostatistical account of Christopher Boorse has shown the naturalistic approach to understanding the concept of disease to be unworkable" (Simon, Carel, and Bird 2017, p. 240). Perhaps more striking is that this conclusion has gone largely unchallenged even by those with considerable knowledge of population biology. Consider, for instance, how the following (otherwise sensible) claim from Sean Valles (2012) digresses: "Quality of life, risk of complications, respect for patient autonomy, etc. are legitimate considerations when deliberating a course of clinical treatment; maximizing fitness is not [...] However, there is a fully-formulated way of closely tying disease and evolutionary insight: the biostatistical theory of health and disease" (p. 256). Valles does not endorse the Bio-Statistical Theory; he is well aware of the criticisms levelled against it. But that he nevertheless positions it as the most respectable *evolutionary* attempt to naturalize biological normativity speaks volumes about its influence. For all its rigor, that account cannot be truthfully characterized as a "fully-formulated way of closely tying together disease and evolutionary insight." The medical literature only very recently began to reflect the theoretical insights of evolutionary medicine (discussed in Section 3). The biological considerations, medical and otherwise, drawn on by Boorse were unavoidably relics of their time and therefore have limited contact with the forefront of evolutionary theorizing today.

2.3 A Hitchhiker's Guide to Naturalizing (Dys)Function: Selected Effects Theory

If the Bio-Statistical Theory was the best that naturalism had to offer, those who sought out non-evolutionary or even nonnaturalistic approaches could be forgiven. Fortunately, it is not. There is another major contender: the "Selected Effects" account of function (Millikan 1984; Neander 1983, 1991a,

1991b; Griffiths 1993; Godfrey-Smith 1993, 1994; Garson 2019). The most recognizable Selected Effects account in the medical (psychiatric) literature is probably that of Jerome Wakefield (1992a), while in the philosophical literature Karen Neander's (1983) and Ruth Millikan's (1984) versions are arguably the most conspicuous. As Millikan explicated the notion of function in exacting and unrivalled detail, I will rely on aspects of her account when discussing the core commitments of the Selected Effects account.

Drawing on the biological notion of teleonomy (Pittendrigh 1958) and Larry Wright's etiological account of function (Wright 1973), the naturalistic account that Millikan developed claims to have three important advantages. First, it explains the existence of a biological component (trait) by appealing to the fitness-enhancing effects that that type of trait provided for ancestors who exhibited it. Second, it grounds biological normativity (i.e., a trait's "Proper function") in a causal history of selective evolution. Doing so importantly sanctions the exclusion of accidental effects or by-products. Third, as a corollary of the second feature, her account provides us with a principled way to determine whether a trait is malfunctioning that does not bow to broad social norms or conventional values. How does her view manage to accomplish what eluded so many others?

Millikan's account employs a distinction between what she calls "first-order reproductively established families" and "second-order (or 'higher-order') reproductively established families." Restricting focus to biological traits and processes, we can make this distinction more comprehensible by substituting less cumbersome terminology. Let us relabel first-order families "genotypes" and higher-order families "phenotypes." A genotypic variant is expressed when the information encoded in the DNA associated with it is used to make protein and RNA molecules. Expression of a genotype contributes to an organism's observable phenotypic traits. Phenotypes are the direct targets of natural selection. Genotypes pass on their structure with very high fidelity but only indirectly because of the differential fitnesses associated with the phenotypes that they produce.

Now, let us apply Millikan's Selected Effects framework to what is by now a too well-worn case: the human heart. We are after a principled way to determine whether a particular person's heart is malfunctioning and thus unhealthy. Judging a biological component dysfunctional presupposes an account of appropriate functioning. Again, for those inclined toward a Naturalistic view of disease, an account of function must be formulated in a way that does not defer to social norms or values when judging some effects desirable and others undesirable. The primary problem is one of how to distinguish among the many distinct causal contributions that the heart makes (or could make) to human fitness. Take, for example, the characteristic "lub-dub" sound made by the heart as its various chambers sequentially contract. It is arguable that this rhythmic sound, not unlike the pumping of blood, is an effect with ramifications for fitness and thus health. When regular, the sound can be an indicator of good health and thus buy its owner peace of mind or enable physicians to readily diagnose a condition that might threaten survival. These distinct causal contributions – pumping blood and making a "lub-dub" sound – are naturally co-instantiated. No mere statistical measure, as suggested by the Bio-Statistical Theory, can rule in favor of one while excluding the other. How, then, can we say of a particular human heart that its proper function is to pump blood rather than make this characteristic sound?

Millikan's resolution, though conceptually demanding, is elegant. Any human heart is a token of a phenotype. The complex genetic-regulatory machinery that reliably produces this phenotype would then be considered the genotype. Millikan argues that the "*Proper*" function of the heart is to pump blood within a designated pressure range because the overall effect of so pumping in some of our ancestors is what gave them a selective advantage over conspecific competitors who had hearts (or heart-like organs) that either failed to pump within that range or did so less reliably. Over time, the selective advantage that accrued to individuals exhibiting this phenotype increased the relative representation of the phenotype. Individuals with such hearts were more likely to survive and produce offspring with similarly performing hearts. Increases in the relative representation of this phenotype necessarily increased the relative representation of the genotypes that reliably produce it. Any presently existing heart is a token instance of a phenotype whose ancestral effects initiated an unbroken causal chain of genotype reproduction. The functional performance of any current heart is thus to be judged against the ancestral standard that explains the heart's existence.

The causal-historical character of this explanation implies a counterfactual claim: were it not for the ancestral (relative) fitness-enhancing effects of pumping blood within a designated range, neither hearts nor the humans that they reside in would be around as we now know them. There is no comparable evolutionary story about the heart's "lub-dub" sound. A corresponding counterfactual claim about the heart's sound would not hold true, or at most appears much less probable. Could we in principle, even if not in practice, disentangle the heart's ability to pump blood from the sound that it happens to make, we would be able to determine that it is only the heart's capacity to pump blood that features as explanatorily necessary. The sound would then be revealed as an incidental by-product of pumping within contingent structural constraints. Fortunately, in this case we need not resort to intuition-mongering or embark on a far-fetched flight through possible worlds to settle the issue. Defective human hearts have long since been replaced with artificial hearts that stem the gap until a donor heart becomes available. Some of the more recently designed "bridges to transplant" are axial flow pumps that whir like little propellers while continuously pushing blood through the body at a constant rate. These devices leave transplant recipients alive but without a pulse (Wieselthaler et al. 2000). The absence of a pulse means that there is no characteristic "lub-dub" sound. Yet the recipient's cardiovascular integrity remains largely uncompromised until transplant. In this explanatory context, the heart's sound is nothing more than noise.

2.4 Hybridizing the Selected Effects Account

Within the philosophy of medicine, Wakefield has arguably been the most ardent defender of the Selected Effects account.[10] Although explicitly addressing concerns in the field of psychiatry, Wakefield's view has implications for the analysis of health and disease more generally. It proposes a resolution to the fact-value debate introduced in Section 1. The lynchpin in his hybrid account is the claim that disorders are *harmful dysfunctions*:

> I propose a hybrid account of disorder as harmful dysfunction, wherein *dysfunction* is a scientific and factual term based in evolutionary biology that refers to the failure of an internal mechanism to perform a natural function for which it was designed, and *harmful* is a value term referring to the consequences that occur to the person because of the dysfunction and are deemed negative by sociocultural standards (1992a, p. 374).

Wakefield's view accommodates Millikan and Neander's Selected Effects account of dysfunction as a necessary condition. His notion of dysfunction, like theirs, is clearly anchored in evolutionary design. However, the philosophical version of SE does not suffice as an account of *disorder* for him. A sufficient account of disorder also requires that a dysfunction cause harm. This is an unambiguously evaluative condition. Taken in conjunction, the descriptive and evaluative conditions are what make his view a hybrid conception of health and disease.

---

[10] Neander's well-known but unpublished dissertation (1983) is also noteworthy for its focus on psychiatry.

What motivates Wakefield's inclusion of an evaluative condition? For him and many others, it is the possibility of conditions that are considered healthy despite reducing individual fitness. A good example comes from life history theory (further discussed in Section 4), which is foundational for evolutionary ecology. Life history theory attempts to explain patterns of intraspecific and interspecific variation in survival (maintenance), growth, and reproductive strategies on the premise that organisms face "trade-offs" arising from energetic, physiological, developmental, or genetic constraints (Stearns 1992). One such fundamental trade-off occurs between reproduction and survival. Selection for maximizing one of these basic proxies for fitness must come at the expense of the other. A particular instance of this trade-off involves traits that increase longevity in women after menopause. Natural selection tends to favor antagonistic pleiotropic variations that increase fertility at the expense of longevity. Traits that increase longevity beyond menopause might accordingly be considered *dysfunctional* from an evolutionary standpoint. While such a variation would reduce reproductive fitness, many would dispute any corresponding evaluation of it as "disordered" or "unhealthy." To the contrary, most of us would probably like to live longer. The practice of medicine consequently follows suit. It aims to mitigate what we see as the unhealthy ramifications (the so-called "diseases of old age") of *non-dysfunctional* variations selected for optimal reproductive effort. There can, it seems, be dysfunction without medical disorder.

The harm condition accordingly plays a crucial role when it comes to licensing attributions of disease for Wakefield and his followers. There is unarguably an element of social choice when it comes to determining harm. The aforementioned example reveals a collective decision to bestow value on a capacity (for longevity) that tends to decrease biological fitness. Medical practice accordingly focuses on promoting healthy longevity and often considers conditions that limit the length or quality of life "disordered." In Sections 4 and 5, we shall examine the details of this evaluative condition more closely. For the time being, however, readers should begin to ponder whether this medical focus on promoting longevity actually contravenes fitness and is, therefore, purely a matter of (non-epistemic) social values.

2.5 Hybridity: Is It the Best of Both Worlds?

The challenge for any hybrid account is whether it can avoid the conceptual weaknesses of the views that it hybridizes while retaining their strengths.

Many have argued that Wakefield's harmful dysfunction account fundamentally collapses into what I have called a wholly Normativist (SNR) position. The legitimacy of this claim will be examined at length in Section 5. For now, it should be noted that there are two very different argumentative routes to such a purported collapse. The first route is, by now, familiar. It holds that admitting even a semblance of normativity or value-ladenness into one's account, as Wakefield's harm condition clearly does, entails a lack of objectivity (Cooper 2002). The arguments presented in Section 1 demonstrate that this inference is suspect. Those who take the second route can agree with my critical assessment of the first route. Still, they might argue that the evidence for and theorizing about evolutionary dysfunction prove to be inconclusive or just hopelessly fraught with difficulties. Little good can come, they might maintain, from a principled (evolutionary) norm that seemingly resists quantification in practice. If this objection cannot be rebutted, the only remaining alternative is Normativism. Along this latter route, one might encounter a motley crew that includes biologists who are staunchly critical of sociobiological approaches to human behavior as well as philosophers of medicine steeped in Continental philosophy. As these may seem like strange bedfellows, I will bring this section to a close by drawing attention to the sort of worry that motivates sophisticated thinkers of very different inclinations.

In its original formulation, the harmful dysfunction account emphasized biological dysfunctions of two types. Wakefield, in the quote presented earlier, refers to the first type of evolutionary dysfunction as the "failure of an internal mechanism." This captures the familiar intuition that disorders occur when "something inside is broken." The other type of evolutionary dysfunction is commonly labeled an "environmental mismatch" (Williams and Nesse 1991). There is, for example, a severe mismatch between the "environment of evolutionary adaptedness" (Bowlby 1969) and our now common obesogenic environments. The preference for and overconsumption of fatty or sugary foods was likely adaptive for our hominid ancestors since opportunities to exploit such energy-dense resources were probably rare. When energy-dense environments are rare, there are few problems with this response; it can be useful in cycles of feast and famine. However, energy-dense foods have been readily available in most developed countries since the agricultural revolution. In most modern societies this exploitative behavioral strategy no longer maintains the fitness advantage that it once held. We are now faced with a situation in which obesity and metabolic disorder have become global epidemics (Gluckman et al. 2005).

Some have questioned whether the foregoing explanations of dysfunction in terms of environmental mismatch or mechanism failure prove adequate. Justin Garson (2021) captures the basic worry:

> What if some of our devastating psychiatric ailments, such as major depression, anxiety disorders, psychopathy, and so on, actually benefited our Pleistocene ancestors? What if, moreover, the *fact* that they benefited those ancestors partly explains why they are around today? Then, if we accept the selected effects theory of function, we would have to say that those disorders do not arise from "dysfunctions." They would be adaptations. (pp. 342–343)

Garson's immediate worry is whether the harmful dysfunction account will pass muster when it comes to nosology and diagnosis in psychiatry, the field supposedly most germane to Wakefield's account. But a more sweeping concern is also apparent. There are few contemporary philosophers who can rival Garson's zeal for the Selected Effects account of function (see Garson 2019). For him to express such grave concerns about the *dysfunction* condition of Wakefield's account is then no small matter.

Wakefield's harm condition was introduced to meet the objection that evolutionary dysfunction alone seems insufficient for attributions of medical disorder. The worry raised by Garson instead questions the necessity of evolutionary dysfunction given the possibility of seemingly disordered conditions that were once adaptive. When taken together, these objections could be taken to show that (Wakefield's acceptance of) the Selected Effects account of dysfunction is neither necessary nor sufficient. If this is so, a fully fledged Normativist view of disorder seems to be the only tenable alternative by default. Some may see fit to conclude as much. I count myself among those who think this a mistake. The next section delves more deeply into the claim that determinations of disorder can make do without a dysfunction condition. Readers will soon see that the evolutionary theorizing that currently informs medical theorizing has come a long way since Wakefield, Neander, and Millikan (among others) made their seminal contributions.

3 Medicine *Sans* Evolutionary Biology?

3.1 Embedding Medicine in an Evolutionary Framework

The examination of biological functionality through an evolutionary lens presents a promising avenue for further philosophical research into the boundaries of health and disease. It provides an objective, naturalistic sense of normativity (teleonomy) for the pivotal distinction between function and dysfunction. But an evolutionary conception of (dys)function must do more than demonstrate its philosophical merits; it should also influence the way medical science perceives disease. We must accordingly take a closer look at how evolutionary biology has changed the prevailing conceptual framework for medicine. In what follows, I delve into research under the label "evolutionary medicine" to determine whether it can change the outlook of traditional medicine. As this section and the next will show, the explanatory framework provided by evolutionary medicine can consolidate the conceptions of function that are often used to argue against the importance of evolutionary theorizing.

3.2 The Dawn of Evolutionary Medicine

While it is probably hyperbolic to pronounce that "Nothing in biology makes sense except in the light of evolutionary biology" (Dobzhansky 1973), one might still hold that "Evolutionary biology is, of course, the scientific foundation for all biology, and biology is the foundation for all medicine" (Nesse and Williams 1998). These words echo the sentiment expressed in a landmark article by the same two authors titled "The Dawn of Darwinian Medicine" (Williams and Nesse 1991) written nearly a decade earlier. Around the time of its publication, they, among others (Ewald 1994), saw the need to remind medical practitioners that, no less than any other type of organism, humans are the outcome of a prolonged process of evolution via natural selection.

The conceptual foundation for this approach has been established under the heading "Darwinian medicine" or "evolutionary medicine" (Nesse and Williams 1994; Perlman 2013; Gluckman et al. 2016; Stearns and Medzhitov 2015). As originally conceived by Williams and Nesse (1991, p. 2), Darwinian medicine's novelty lies primarily in what it offers to our understanding of disease causation. They singled out four categories of such causation. The first category is that of infection. Evolutionary theorizing has immediate relevance because it sheds light on the dynamics of "conflicts between pathogens and their human hosts as well as the adaptations with which both contestants attempt to influence the outcome in their own favor." They suggest that "an evolutionary taxonomy of manifestations of infectious disease must underlie any attempt to understand such conflicts, and that an appreciation of the implications of the rapid evolution of pathogens may help with the solution of some public health problems." Their second category involves the "understanding and treatment of physical injuries that result from mechanical or chemical agents." An evolutionary perspective emphasizes the value of distinguishing among several kinds of repair mechanisms (e.g., rebuilding damaged tissues) and secondary processes of adjustment (e.g., increases in temperature associated with inflammation) that indirectly aid in repair. Evolutionary theorizing also promises to contribute to a third category concerning "genetic disease, and genetic factors that affect susceptibility to disease, especially those maintained by pleiotropic effects that offer benefits in youth but exact a cost later in life by increasing susceptibility to the diseases of aging." Their final category involves examining "how differences between present circumstances and the environment of evolutionary adaptedness may contribute to diseases of civilization including deficiency states, toxins from various natural and artificial sources, and some behavioral and psychological disorders."

In all four categories, Williams and Nesse show how evolutionary theorizing compels medicine to consider causal factors other than the biomechanical and biochemical ones that traditionally feature in physiology or development. This, they argue, poses a fundamental challenge to traditional medical conceptions of disease:

> Disease looks different from an evolutionary perspective. Infection is not a happenstance encounter with another organism, but an arms race between host and parasite, with extraordinary elaborations of weapons, strategies, defenses and counterdefenses. Trauma is not a mere matter of damaged tissue, but of the failure of protective mechanisms, the yielding of the soma at weak spots, and repair processes that have been shaped and constrained by natural selection. Genes that cause disease are not just the result of mutation, but may be selected for known or unknown benefits, such as the vigor in youth that may result from genes that later cause aging. Environmental abnormalities, not limited to changes in the last few generations, are major causes of common diseases, often in interaction with genetic "quirks" that are harmless in the environment of evolutionary adaptedness. (Op cit. pp. 16–17)

The message to take away from this challenge is not that physiological or developmental explanations of such phenomena are wrong. Nor is it that phylogenetic and selective explanations provide better answers. Rather, their central point is that physiology and development aim to answer different questions and thereby provide only a partial explanation of the phenomenon of disease.

Coming to grips with this framework and its growing importance requires recognizing that the evolutionary considerations it introduces are characterized rather broadly. In fact, the "evolutionary" considerations just mentioned plainly include what would more commonly be described as *ecological* factors. Inquiries into evolutionary history (e.g., phylogenetics, comparative morphology, and evolutionary developmental biology) do not exhaust the evolutionary component of evolutionary medicine. Such inquiries are supplemented by applications of population genetics, quantitative genetics, behavioral ecology, population ecology, and even community ecology (e.g., microbiome research) to contemporary populations. Noting this, Pierre-Olivier Méthot (2011) has argued that Darwinian medicine and evolutionary medicine should be seen as distinct "research traditions" or "orientations":

> Whereas [evolutionary medicine] primarily involves "forward looking" explanations, [Darwinian medicine] depends mostly on "backward looking" explanations. A "forward looking" explanation tries to predict the effects of ongoing evolutionary processes on human health and disease in contemporary environments (e.g., hospitals). In contrast, a "backward looking" explanation typically applies evolutionary principles from the vantage point of the evolutionary past of humans (here, the Pleistocene epoch) in order to assess present states of health and disease among populations. The contrast between these two explanatory styles can also be captured by the distinction between a theoretically and a practically oriented approach; whereas evolutionary medicine seeks to devise practical solutions to medical problems based on specific applications of evolutionary biology's toolbox, Darwinian medicine, in contrast, stresses the need to compare past and present populations from an evolutionary point of view in order to gain insights into why we in the present get sick. *Both approaches, however, are ultimately concerned with the prevention and control of human diseases* (pp. 76–77; my emphasis).

While Méthot's distinction has gained considerable traction among philosophers, let us momentarily collapse the distinction between evolutionary medicine and Darwinian medicine based on their shared "ultimate concern" (italicized in the quote above). Doing so will ease exposition and aid comprehension in this section. There are principled as well as practical reasons for collapsing it, ones that shall be explained in Section 4. In the meantime, however, I designate this expanded, undifferentiated program Evolutionary Medicine (uppercase "E" and "M") to distinguish it from Méthot's more restrictive notion.

The most important insight of Evolutionary Medicine is that it augments the usual focus of medical research on the mechanisms of disease by asking why the body is constructed the way it is. Modern medical research typically takes the structure of the body as a given and consequently focuses on questions about physiological mechanisms and how they differ in individuals with disease. The logic behind this approach is not much changed from the one codified in John Stuart Mill's (1846 [1865]) informal (i.e., inductive) methods for inferring causal relationships. By way of example, consider Mill's "joint method," which combines his methods of agreement and difference. One is thereby enjoined to compare a variety of situations in which a suspected factor is present to similar situations in which that factor is absent and then show that a certain effect is observed in all and only those instances in which the suspected factor is present. While inferential methods have certainly become more sophisticated (Rohlf and Sokal 1995; Woodward 2003), the basic diagnostic strategy in much of medicine still involves working out the relevant causal *differences* between those who are healthy and those who are not. The aim is to answer "proximate" questions (Mayr 1961) about how the human body works and what goes awry in contrast classes consisting of supposedly diseased subjects. While this approach remains indispensable to modern medicine, it is incomplete. Evolutionary Medicine complements this prevailing strategy by redirecting the focus of research. It asks why all individuals are alike in ways that leave them susceptible to disease and then asks how evolutionary factors might account for such vulnerabilities (Williams and Nesse 1991; Nesse and Williams 1998). Evolutionary Medicine thereby emphasizes *similarities* rather than differences. These "ultimate" evolutionary questions must also be answered in order to have a full understanding of the body and disease.

None of the foregoing should be taken as suggesting that modern medicine *never* concerns itself with similarities. Neither should it be taken as a claim to the effect that Evolutionary Medicine considers only similarities at the expense of differences. It is more a matter of emphasis. Turner syndrome, for example, is a condition that affects only females. The disorder is the result of a completely or partially missing sex chromosome. It can cause a slew of medical and developmental problems, including heart defect, short stature, and developmental failure of the ovaries. Girls and women with Turner syndrome need ongoing medical care from a variety of multidisciplinary specialists. If they receive appropriate care, many girls and women can lead relatively healthy and independent lives. Contemporary medical practice obviously recognizes the importance of a *similarity* (biological sex) among those with this condition prior to any attempted management via growth hormone therapy and hormone replacement therapy. However, this similarity is very much a latent background assumption for subsequent medical research and intervention. While Evolutionary Medicine tends to emphasize similarity, it does not discount the importance of *differences*. The similarities identified in one demographic group (e.g., females) or even a species (e.g., *Homo sapiens*) can be recharacterized as a relevant difference when explanatory focus encompasses other groups (females vs. males) or even species (comparative explanations involving other primates). In fact, it is precisely by virtue of these differences (within or between populations, and those among species) that researchers are able to identify plausible ranges of functionality for those bodily parts and processes whose ancestral states and ranges are inaccessible to direct measurement.

Basic examples can help further illuminate the importance of evolutionary considerations. Consider, first, the deceptively simple case of fever (pyrexia). To be diagnosed with a fever one must have an elevated body temperature above a normal range due to an increase in the body's temperature set point. Exhibiting a fever can indicate the presence of many underlying conditions, ranging from viral or bacterial infections to cancer. Having a fever is atypical and discomforting for most. But a fever is not thereby a defect or disease. To use Williams and Nesse's (1991, p. 9) terminology, it is a "secondary response" or "process that indirectly aids in repair" and thus an evolved defense of the human immune system. Immunological responses typically facilitate the elimination of pathogens or diseases that cause them. Fever, in particular, initiates a cascading series of cellular mechanisms (i.e., the febrile response) that regulate the expression of genes whose protein products determine the appropriate level of immune system response to infection or disease (Harper et al. 2018). Still, some physicians unhesitatingly administer fever reducing drugs (antipyretics) to alleviate the discomfort experienced by a fever-ridden patient. An evolutionary perspective reveals the potential harm in this well-intentioned Hippocratic reflex to immediately alleviate distress. Evolved adaptive responses need be neither comforting nor desirable for those who experience them. What matters, from a purely evolutionary standpoint, is whether such responses increased the probability of survival and reproduction over a lifetime for our ancestors. Had we the means to reliably eliminate fever without ever compromising any of the benefits of an effective immune response, we would not hesitate to avail ourselves of such means. But we don't always have this luxury. Unless a fever becomes life-threateningly high or introduces other complications, fever reduction should be a carefully considered decision that carefully balances a physician's (at times competing) duties to alleviate patient distress and do no harm.[11] Alleviating fever has the potential to prolong the duration of acute

---

[11] It is noteworthy that the febrile response can also be hijacked by pathogens. The periodic spiking fever of patients with malaria almost certainly represents manipulation of the

illness. An evolutionary understanding of the adaptive benefits of fever and the associated complications that antipyretic therapy might introduce reveals this in a way that has profound implications for whether and how to administer treatment.

Another instructive example involves cystic fibrosis (hereafter "CF"). It is caused by an autosomal recessive mutation in the CFTR (cystic fibrosis transmembrane conductance regulator) gene that codes for a chloride channel. Individuals who are homozygous recessive have cells with defective chloride channels that cannot decrease the viscosity of mucus secretions. Mucus in CF patients is thus very thick and tends to accumulate in the intestines and lung, where reduced ciliary mucus clearance then increases susceptibility to pulmonary infections that eventuate in respiratory failure. Breathing difficulties and respiratory infections aside, CF patients also often suffer from malnutrition and exhibit poor growth, though lung disease remains the most common cause of death. Despite a high juvenile fatality rate and the fact that almost all (97–98%) males with CF are infertile, there is an increased prevalence of CF in those of Caucasian or European ancestry compared to those of African descent. The relative prevalence of newborn "carriers" in those of Caucasian-European ancestry (e.g., ~1:2,000 in the United Kingdom vs. ~1:29,000 for African Americans) is also markedly pronounced (Klinger 1994). These observations cry out for explanation because selection usually eliminates seriously deleterious conditions that are caused by a single gene alteration. Although the reasons for it still remain unclear, it has been hypothesized that this higher relative prevalence may have been initiated during outbreaks of either cholera or (more likely) tuberculosis in Europe.[12] Evidence suggests that being a heterozygous carrier could have conferred a selective advantage upon affected individuals during these outbreaks (Mowat 2017). Cholera, for example, produces a toxin that hyperactivates the chloride channels affected by mutations in the CFTR gene. This direct hyperactivation could have caused historically lethal bouts of intense diarrhea. Because CF carriers have fewer (about half as many) functioning chloride channels, their cells are less prone to the efflux of ions that cause diarrhea (Withrock et al. 2015). Periodic outbreaks of tuberculosis present a slightly different scenario in that the selective advantage of being a carrier

---

patients' immune systems for the benefit of the pathogen. In this situation, treatment of fever can actually be justified *on evolutionary grounds*. I thank an anonymous reviewer for bring this example to my attention.

[12] There is currently no consensus in the cystic fibrosis research community. However, see Mowat (2017) for reasons to think that the tuberculosis hypothesis is more promising.

is supposedly associated with the presence of an environmental stressor. It is speculated that the domestication of cattle by hunter-gatherers, who turned to farming as they slowly migrated north (into Europe), required a period of adaption to overcome lactose intolerance. While acclimating to a new diet including milk, these migrants would have experienced unavoidable lactose-induced diarrhea and resulting dehydration. Here, too, one can see how fewer functioning chloride channels having might have been an advantage (Mowat 2017, p. 169). In either case, there appears to be a direct causal link between Caucasian people and the selective advantage (resistance to severe diarrhea/dehydration) that comes with being a carrier. Carriers were more likely to survive such outbreaks than those who did not carry the recessive deleterious allele. They were subsequently more likely to mate with individuals who were also carriers. But when carriers mate with carriers, there is always a chance that their offspring will exhibit either the homozygous recessive CF phenotype (25%) or be heterozygous carriers themselves (50%). There is accordingly a 75% chance that their offspring will have at least one copy of this lethal allele. It is no wonder that the prevalence of CF in those of Caucasian-European descent is orders of magnitude more common than what we would expect for such a deleterious condition.[13] Observed differences in the prevalence of CF would be an inexplicable medical mystery in the absence of an evolution-infused explanatory framework.

Evolutionary models can and have been used to understand the persistence of disease-causing genes with fitness-diminishing effects. Other applications of evolutionary theory, such as the implications of maternal-fetal conflict for cancer research (Haig 2015) or how predictive adaptive response theory informs our understanding of metabolic syndrome (Bateson et al. 2004), are more challenging. Given just the examples above, however, one might suppose that evolutionary biology regularly contributes to the scientific understanding of disease in ways that can inform the treatment of patients. If you count yourself among those who feel this way, it may come as a surprise that there remains significant opposition to the use of evolutionary theory in medicine.

---

[13] Readers may find it strange to think of this condition or the recessive deleterious allele as being in any way "common." In an absolute sense, the deleterious recessive allele for CF is uncommon, generally having a prevalence rate of, for example, 0.04 in the United Kingdom. However, this reveals that it is actually quite common *relative to* the thousands of other heritable states that are known to be harmful to fitness. The case of Huntington's Disease, an autosomal dominant disorder, presents a good contrast. As of 2010, its prevalence in the UK was approximately 0.00012 (Wexler et al. 2016).

3.3 The Persistent Resistance to Evolutionary Medicine

The last three decades have seen a steadily growing appreciation of the contributions that Evolutionary Medicine can make. But it still remains largely on the margins of mainstream medical practice. This is a somewhat perplexing state of affairs when one considers the important contributions of evolutionary theorizing in other areas of human concern, notably agricultural settings featuring animal husbandry (selective breeding) and the evolution of pesticide resistance. Its marginalization becomes even more difficult to comprehend when one remembers that a great deal of early evolutionary theorizing was done by (medical) doctors. Charles Darwin's grandfather, Erasmus Darwin, is a case in point. The medical interests of Erasmus were steeped in evolutionary considerations: "The purport of the following pages [of *Zoonomia, or the Laws of Organic Life* (1794/1809)] is an endeavour to reduce the facts belonging to animal life into classes, orders, genre and species; and by comparing with each other to unravel the theory of disease." What tore apart this seemingly happy marriage of evolutionary theory to medicine?

A history of the sometimes-tortured relationship between medicine and biological theorizing in the nineteenth century would of course be informative on this score. However, limitations of space preclude a summary of that fascinating history here.[14] Discussion will be restricted to two more recent lines of resistance to evolutionary theorizing in medicine. I will refer to resistance in the first sense as founded on "biological considerations," while resistance in the second sense has its origin in what I will call "the pragmatic contention." As a good deal of scholarly work has already been devoted to the first kind of resistance, I will discuss it somewhat more quickly. The second contention has influenced the philosophy of medicine in important ways that too often go unrecognized. It thus merits more careful consideration.

The criticisms of applied evolutionary theorizing that fall under the heading "biological considerations" took discernable shape shortly after the crystallization of sociobiology (Wilson 1975).[15] Sociobiological inquiry in general aims to explain social organization and behavior via evolutionary principles, often by appeal to selective explanations. In many ways, it

---

[14] For readers interested in the history of this relationship, I recommend Buklijas and Gluckman (2013) as well as Zampieri (2009).

[15] The story is more complicated than this statement suggests. The history of eugenics and social Darwinism cannot be ignored in this context. For present purposes, however, beginning with E. O. Wilson's work introduces no serious distortions.

and its subsequent manifestations (Barkow, Cosmides, and Tooby 1992; Davies and Krebs 1984; Boyd and Richerson 1988) have paved the way for evolutionary medicine. Critics have taken issue with sociobiological research on scientific and moral grounds, often likening its uptake to the worst of eugenicist thinking (Sociobiology Study Group of Science for the People 1976).

While some of this criticism was undoubtedly motivated by political disagreements (Segerstrale 2000), it also raised important methodological considerations about the assumptions that guide sociobiological research. In particular, critics of sociobiology cautioned against "adaptationist thinking" of the form that takes natural selection to be the predominant if not exclusive causal factor at work in the establishment and maintenance of traits. Peter Godfrey-Smith (2001) has dubbed this particularly fervent form of adaptationist thinking "Empirical Adaptationism." Perhaps the single most influential work in a critical mode was penned by the Harvard University biologists Richard Lewontin and Stephen J. Gould. Their article, "The Spandrels of San Marco and the Panglossian Paradigm: A Critique of the Adaptationist Programme" (1979), introduces a host of (mostly) sensible reservations draped in provocative literary references and scathing satire. They depicted rampant adaptationist thinking as hopelessly simple and optimistic ("Panglossian" following Voltaire 1759) by highlighting the evolutionary role of nonselective factors, notably drift and developmental correlation (e.g., allometry). Ignoring or even underestimating such well-documented factors, they argued, can lead to the hasty conclusion that most traits exist in the form that they do because of active (past or present) selection for fitness-enhancing effects.

These biological considerations have considerable ramifications for evolutionary medicine. Evolutionary Medicine in its most general form examines the human conditions that lend to susceptibility and physical distress. One of its pivotal insights is that even the unpleasant can sometimes be adaptive, as we often see with immunological responses (e.g., fever). But when there is no readily available adaptive explanation of discomfort in terms of *current* fitness enhancement, Evolutionary Medicine typically shifts to explaining why fitness-diminishing conditions are observed in current populations to the extent that they are (e.g., CF in Caucasians). This shift does not necessarily exclude consideration of nonselective explanatory factors such as rates of mutation or migration and drift. Evolutionary Medicine does not, then, fall prey to many of the allegations that have been levied against naïve selective explanations. More often than not, however, another defeasible selective hypothesis is introduced and explored before conceding a reduced explanatory role for natural selection. This repeated introduction and testing of modified selective hypotheses is precisely what defines the commitment to "Methodological Adaptationism" (Godfrey-Smith 2001). An issue of concern for philosophers of biomedicine who remain committed to evolutionary theory is to clarify when maintaining this commitment to selective explanations of disease as dysfunction no longer proves worthwhile (Lewens 2015).

The second important strand of resistance to Evolutionary Medicine emerges from a pragmatic contention. The utility at issue here is one which commonly pertains to general practitioners, family doctors, and even specialists on the front lines of medical practice. Setting aside routine health checks (preventative visits), they are usually confronted with patients experiencing distress. Patients desire remedies that alleviate their symptoms effectively, affordably, and as quickly as possible. Navigating these potentially competing criteria requires compromise. What a patient desires must be pitted against what a physician deems proper given the patient's condition, not to mention what is considered best practice from a public health standpoint. Still, the most immediate pressure a doctor experiences comes from a patient's request for help and the Hippocratic Oath (or a modern equivalent) that compels assistance. This motivational pressure, perhaps more so than any other, can occasionally drive physicians to make what appear to be rash decisions about the proper course of treatment, ones that defy their overarching directive to do no harm.

The overprescription of antibiotics presents a familiar example in this context. Physicians have often prescribed antibiotics even when they suspect that doing so may not affect a patient's prospects for recovery (Wang et al. 1999), as would be the case for illnesses with viral rather than bacterial etiologies. Some physicians may rationalize this decision on the grounds that it is a "low-risk" strategy when considering the health outcomes for a single patient. Even if a patient's condition is not caused by a bacterial agent, administering antibiotics typically does not diminish the patient's immune response to a nonbacterial pathogen. If the patient is likely to recover with or without antibiotics, the only sure cost is then a financial one to the patient's or perhaps an insurer's pocketbook. Moreover, the fact that a patient has been prescribed something that she believes could aid in recovery will usually make her feel better about the current situation. Administering antibiotics might thus have a placebo effect. Even if we rule out the possibility of placebo effects, patients are frequently more satisfied with a physician who actively does something (commits) rather than nothing (omits).

Unfortunately, it is now well known that there are serious drawbacks to prescribing antibiotics haphazardly. Bacterial populations exposed to antibiotics show strong selection for resistance. Since bacterial population sizes are extremely large, it is likely that such populations already include genetic variations that induce resistance. We can put to rest any doubt about this just by noting that "[a] single gram of fecal matter [in the human small intestine] is likely to include at least one newly occurred instance of every single point mutation" for any bacterial strain it happens to host (Genereux and Bergstrom 2005). Bacteria also have very short generation times. Rapid reproduction, ample variation, and strong selection in antibiotic-infused environments virtually guarantee that mutations for resistance will arise and sweep to fixation at an alarming pace.

The evolution of antibiotic resistance counsels against the pragmatic contention depicted earlier. Some physicians are nowadays considerably more hesitant about prescribing antibiotics, thanks in large part to better communication and understanding of evolutionary insights (e.g., "antimicrobial stewardship" guidelines). Nevertheless, for most physicians, basic medical care apparently does not require a great deal of thought about the underlying science. Gilbert Omenn, a medical doctor and researcher at the University of Michigan, confirms as much: "The why and even the how is not essential, if you have good published evidence that something works and you've seen it work in some of your patients, then it's enough to try and help your patient as best you can" (Diep 2017). Omenn is not alone in thinking this way. What I call "the pragmatic contention" finds a home even among those who perform procedures in more specialized medical contexts. David Gorski, a surgeon and researcher at Wayne State University, expresses much the same sentiment: "Most physicians are not scientists. This is not a knock, but they're more akin to engineers [...] They take science that's already known and they apply it to a problem, the problem being making patients better [...] To be honest, to do an operation, you probably don't need to understand evolution" (Diep 2017). The general impression that much of medicine can make do without regard for the insights of evolutionary theorizing has also been reinforced by the way aspiring medical practitioners are typically educated. Until very recently, the biological coursework required for medical credentialing, both pre-med and during medical school, included evolutionary biology only as an afterthought (Nesse and Schiffman 2003; Downie 2004; Alcock and Schwartz 2011).

Notice that Gorski's terminology ("you *probably* don't need to know *a lot* about evolution") hints at a subtle and challenging shift. There is a whiff of prudent hesitancy on his part. A *great deal* of evolutionary theory is not *obviously* necessary for some aspects of medicine. This is compatible with there being a lot of evolutionary theory that is inexplicit because it is taken for granted as obvious, or there being just a small but indispensable amount of evolutionary theory that informs diagnosis, prognosis, and intervention. Either way, it is a mistake to suppose that "the pragmatic contention" makes evolutionary theorizing completely irrelevant or even unimportant to the practice of medicine. Evolution may not be at the very forefront of a surgeon's mind when performing an operation. Mechanistic understanding of the body and procedural knowledge often appear adequate. But appearances can be deceiving. The surgeons who operate must also prescribe antibiotics to prevent post-surgical infection. Prescribing and correctly administering antibiotics are essential medical interventions that are no less important than surgery itself. Furthermore, closely examining why surgeons perform particular corrective procedures in the first place or in the specific ways that they do can often disclose the subtle influence of evolutionary theorizing. Contrast this with the case of general practitioners who incautiously prescribe antibiotics. They *mistakenly* ignore or downplay the relevance of Evolutionary Medicine. This mistake could be due to inadequate training or effort. Championing individual (patient) well-being at the expense of public health is another, somewhat more nefarious, possibility. No matter what the reason, the consequences of ignorance or deemphasis are unacceptable. We correspondingly attempt to change prevailing opinions and actions, usually via educational or punitive measures. It is considerably more difficult to demonstrate that good surgeons are somehow negligent in their choice to ignore or downplay the insights of Evolutionary Medicine. In fact, it seems plainly wrong to call their pragmatically motivated oversight "antievolutionary." It would perhaps be better described as a form of methodological indifference or localized apathy.

"The pragmatic contention" is not just a widely held intuition among medical practitioners. Longstanding philosophical disputes about the proper analysis of function and what it is to malfunction can also turn on pragmatic considerations related to sub-disciplinary practice within biology (Amundson and Lauder 1994). In spite of this, many have steadfastly summoned the resources of evolutionary biology to ground the notion of (dys)function in a way that might legitimize a distinction between the healthy and the unhealthy. However, there remain many who question the wisdom behind this move, not all of whom are ignorant of or completely averse to the incursions of evolutionary theorizing. Tim Lewens, for instance, has expressed his skepticism when noting that "the shift to an evolutionary view makes pathology hostage to evolutionary enquiry more generally" (2015, p. 188).

In the next section, I will assess whether this consequence is as objectionable as Lewens and others imply. We have already seen how a general evolutionary account of (dys)function might prove philosophically useful. The current section shelved that philosophical debate and introduced some of the advantages that seemingly accrue to conventional medicine when it takes evolutionary theory seriously. But we have only just scratched the surface of the conceptual edifice erected by Evolutionary Medicine. We must now examine how it has managed to sophisticate the dysfunction condition as traditionally articulated by Selected Effects and hybrid theorists. Section 4 accordingly delves more deeply into the philosophical utility of thinking along the lines suggested by Evolutionary Medicine.

4 The Biology of Disorder: A Burgeoning Philosophical Consensus

4.1 Building a Better Dysfunction Condition via Evolutionary Medicine

Promissory notes come with expectations of fulfillment. To this point, I have detailed some of the limitations of a purely Normativist (SNC) account of pathology and hinted at the inadequacies of prevailing Naturalistic (ONR) counterparts that attempt to naturalize dysfunction. Those difficulties certainly make a hybrid account more appealing. The first impediment to a cogent hybrid account is not so much the imposition of normative considerations (i.e., the "harm condition"), which will be addressed in Section 5, as it is the limitations of the biological dysfunction condition as traditionally conceived. The time has arrived to reveal what a more nuanced, scientifically updated version of the dysfunction condition for disorder should look like, and to determine whether it can quell the anxieties of those who would otherwise opt for some form of Normativism.

An initial step in this direction is to admit that critics were right to question the adequacy of the evolutionary dysfunction condition for medical disorders. While oversimplified versions infect nearly all Selected Effects accounts to date,[16] critics have taken particular issue with the rendition of it in Wakefield's account. He argued that a physical condition can be

---

[16] For a notable exception, see Christie et al. (2022).

pathological *only if* there is a "mechanism(s) failure" or an "environmental mismatch." This simple dichotomy has proven too crude. As I will show, it was hindered by a reliance on mid-twentieth century thinking about the way that evolutionary considerations bear on explanations of function.

4.2 Tinbergen's "Four Questions": A Foundation for Evolutionary Medicine and the Philosophy of Biomedicine

In Section 3, readers were introduced to Evolutionary Medicine in a broad sense that deliberately collapsed Méthot's (2011) influential distinction between "evolutionary medicine" and "Darwinian medicine." According to Méthot, the former is *prospective* in that it attempts to predict the effects of ongoing evolutionary processes on human health and disease in contemporary environments, while the latter is *retrospective* insofar as it applies evolutionary principles from the vantage point of the evolutionary past of humans to assess present states of health and disease among populations. It was also argued in Section 3 that the medical insights gleaned from evolutionary considerations become apparent when one distinguishes ultimate from proximate causes (Mayr 1961). *Proximate* explanations emphasize the relevant causal differences between the healthy and the unhealthy. Evolutionary Medicine, in contrast, stresses similarities rather than differences. It asks why all individuals are alike in ways that leave them susceptible to disease and then asks how evolutionary factors might account for such vulnerabilities (Williams and Nesse 1991; Nesse and Williams 1998). These *ultimate* evolutionary questions must also be answered if we are to have a full understanding of the body.

For expository purposes, it proves useful to collapse the distinction between prospective and retrospective evolutionary explanations and focus instead on the proximate–ultimate dichotomy. This strategy has its drawbacks, though. Chief among these is that it can inadvertently promote the misconception that these distinctions always run in parallel. It is a mistake to presume that forward-looking evolutionary medicine never concerns itself with ultimate causes. And it is similarly incorrect to suppose that Darwinian medicine is rarely if ever preoccupied with proximate causes. Any depiction that suggests otherwise is, at best, a caricature of current research in Evolutionary Medicine.

The ground for a more sophisticated understanding of pathology was broken by evolutionary ethologist Niko Tinbergen's "On Aims and Methods of Ethology" (1963). Tinbergen shows that there are actually four distinct questions that one can ask when attempting to understand any biological characteristic. The first question to which he drew attention was one of "causation." Using contemporary parlance, we would say that such questions call for investigation into the mechanisms that underly a trait at a specified time. Readers will by now be familiar with this type of explanatory project from discussion of the Causal Role account of function in Section 2. The second kind of question Tinbergen noted concerns the "survival value" of a trait. Such questions require answers that involve the current or original function of a trait. How, in other words, does (has) the trait in question contribute(d) to the survival and reproduction of the organisms exhibiting it? Here it would perhaps be better to replace Tinbergen's label "survival value" with either "current utility" (Bateson and Laland 2013) or "adaptive significance" (Nesse 2013). The latter two terms better integrate the fact that reproductive fecundity (contra mere viability) is often the currency of trade for natural selection. Yet a third type of question demands detail about how a trait develops over the life-course of individuals. Tinbergen noted that questions of this kind raise issues of "ontogeny" or organismal development. The fourth and final type of question is a plea for what he called "evolutionary" considerations. While this label alone is too vague to be informative, Tinbergen's intent was clearly to emphasize a role for questions that address phylogenetic comparison among closely related species in a clade as well as studying features of fossilized humans. Such comparative analyses can establish whether and how the features of a trait have changed over durations longer than the individual's lifetime.

Tinbergen's "four questions" can be readily organized in a two-by-two table (Table 1):

Table 1 Tinbergen's "Four Questions" (See text for details.)

	Dynamic or Diachronic Target of Explanation (over some duration of time)	**Static or Synchronic Target of Explanation** (at a specified point in time)
Proximate Explanation ("How")	"Ontogeny" (individual development)	"Causation" (intrinsic and extrinsic causal mechanisms)
Ultimate Explanation ("Why")	"Evolution" (phylogeny)	"Survival Value" (functional explanation of current utility or adaptive significance in terms of reproductive capacity)

Of utmost importance is the way that Tinbergen and many others since have seen fit to group these questions. As shown in Table 1, questions pertaining to "ontogeny" (organismal development) and "causation" (mechanistic underpinnings of traits) are supposed to be conceived of as proximate. "Evolutionary" (phylogenetic) questions and the problems posed by "survival value" (functionality) are considered ultimate. This clearly evinces a considerable extension of Mayr's (1961) more basic "proximate vs. ultimate" dichotomy. No less important was Tinbergen's recognition that these are *complementary* explanations. All of these must be undertaken in concert and integrated if we are to have a complete understanding of any biological character.

Tinbergen's framework certainly makes for a substantial conceptual improvement, but it is still not sophisticated enough. Much has transpired since its introduction more than half a century ago. Patrick Bateson and Kevin Lala (formerly Laland) summarize this point succinctly: "Over the past 50 years, major developments have occurred in the understanding of extra-genetic inheritance processes, such as cytoplasmic effects, parental effects, including maternal and paternal genomic imprinting and other epigenetic impacts on gene expression, ecological legacies, behavioural traditions, and cultural inheritance" (2013, p. 713). These conceptual, theoretical, and empirical advances are staggering.[17] I will review but one reason for questioning the adequacy of Tinbergen's framework as originally presented.[18]

Consider explanations that fall into the category he labeled "ontogenetic." Such questions supposedly prompt dynamic, proximate explanations (see Table 1). Whatever the duration of time targeted by an ontogenetic explanation, Tinbergen presumed that it must elapse within the individual's lifetime and thereby reflect an *intra*generational process. Otherwise, there would have been little justification for the label "proximate." Individual development and especially the developmentally plastic responses that characterize human growth through reproductive maturity are sensitive to many aspects of the environment. If extra-genetic processes of inheritance such as parental effects (e.g., maternal imprinting) are taken seriously, then it becomes evident that *inter*generational and potentially *trans*generational

---

[17] Readers should consult Pigliucci and Müller (2010) and Laland et al. (2015) for details about the so-called "Extended Evolutionary Synthesis."

[18] While I question the classic ultimate–proximate distinction primarily from a scientific perspective, it is worth mentioning that some philosophers defend its explanatory utility by appeal to the complementary roles of "structuring" and "triggering" causes (Ramsey and Aaby 2022).

factors may not only influence features of the environment that offspring encounter but also affect the suite or probability of phenotypic responses available to offspring.[19] The predictive adaptive response of offspring metabolism to low-nutrient conditions (e.g., famine) in the parental generation mentioned in Section 1 is an instance of this (Bateson et al. 2004; Gluckman et al. 2005). This is an *inter*generational effect insofar as it is initiated by epigenetic modifications in utero that are caused by the mother's nutritional uptake. The nutritional status of the mother's diet establishes an intrauterine environment that modifies the developing metabolism of offspring for the nutritional environment they are most likely to encounter (cf. Wells 2012). Examples such as this suggest that "proximate" and "ultimate" explanations are not as self-evident or easily distinguished as Tinbergen and others have assumed. These categories may blur into one another along a continuum that is only conventionally discrete.

4.3 Tinbergen's Framework: A Prized Relic for the Philosophy of Biomedicine

In spite of the challenges posed by recent advances in biology, Tinbergen's framework of explanation remains a fairly good heuristic. The community of researchers in Evolutionary Medicine continually draw on it. Its legacy to philosophy of medicine is the provision of more refined objective criteria for judgments of dysfunction. Philosophers of biomedicine have recently exploited the fact that corollaries of Tinbergen's quadripartite division reveal distinct ways that traits can fall short of optimal performance. Drawing on this insight, they have articulated much more nuanced conditions for dysfunction than those proposed by Wakefield or Selected Effects theorists.

No philosophers have done more to reorganize thinking about attributions of pathology than Matthewson and Griffiths (2017, 2018). There are, according to them, "four ways of going wrong" that can in many ways be seen as anticomplementary to what Tinbergen might have considered the "four ways of going right" (i.e., answers to the four types of questions in Table 1).

---

[19] The distinction between "intergenerational" and "transgenerational" is often used in discussions of epigenetic inheritance to distinguish the (intergenerational) influence of a focal progenitor (a biological mother) on its immediate offspring from the (transgenerational) influence it exerts on its grand-offspring or even great grand-offspring.

Two of their four ways will already be familiar to readers from previous sections: "mechanism(s) failure" and "environmental mismatch." These were recognized by the fathers of Darwinian medicine and are of course foundational for Wakefield's dysfunction condition. A mechanism failure occurs when a biological component or subsystem fails to bring about a resultant state of affairs that is necessary to maintain organismal survival and optimize reproductive capacity. Disorders resulting from environmental mismatches presuppose that an organism is intact and undamaged. There are no faulty mechanisms or intrinsic deficiencies. The threat to organismal fitness instead comes in the form of encountering environmental conditions that are detrimentally novel. The class of disorders due to mechanism failure(s) presents us with a pathological analogue of Tinbergen's "causation" (causal mechanisms) category of static, proximate questions. Environmental mismatches correspond most closely to disorders that would arise in the category of "survival value."

The third "way of going wrong" represents a significant departure from more orthodox medical thinking about human dysfunction. Following Matthewson and Griffiths (2017, p. 455), I will refer to such disorders as the result of "normal but inhospitable environments." A "normal" environment here would be one that falls along the actualized range of environments in which a biological trait was historically selected. The measured values of relevant biotic or abiotic factors in a particular normal environment accordingly figure into calculations of the parameter values for a hypothetical average reference environment deployed in models of selection and might thereby correspond roughly with what John Bowlby called the "environment of evolutionary adaptedness" (1969).

It is important to recognize that the conditions that define a normal but inhospitable selective environment could have been statistically uncommon or present considerable challenges to survival. Being infrequent does not make such conditions unimportant from an evolutionary perspective. A generic example can show why. Consider a range of normal environments that happens to include a rare but extremely inhospitable environment. So inhospitable is this relatively uncommon environment that inhabiting it in the absence of an adaptive plastic response (i.e., exhibiting a specialized, but ill-suited, nonplastic alternative) virtually guarantees immediate death. Further assume that a reproductive cost accompanies the capacity for plasticity. An organism could, in other words, produce more offspring if it did not dedicate resources to developing and maintaining this capacity. Despite the fitness cost, there can be positive selection for plasticity and a corresponding increase in its representation. This can occur when the average environment of selection over evolutionary time includes a low but nonnegligible probability of encountering a potentially lethal (for nonplastic variants) normal environment. Other things being equal, the probability of encountering such an environment increases with time.[20] Even a slim chance of encountering this very inhospitable normal environment can generate selection for plasticity over specialization. There may, in effect, be selection for doing less well than one might otherwise have done when the alternative is extinction.

Implicit in the discussion so far is the basis for classifying an environment as "inhospitable." Environments can be inhospitable in countless different ways. The longer or more closely any habitat is examined, the more obvious it becomes that there are irregularities in the quality or quantity of available resources as well as disparities concerning the (demographic) impact of extrinsic factors. Spatially partitioning a *macro*-environment often reveals a patchwork of variable *micro*-environments. Even the more spatially homogenous environments tend to exhibit significant variation over distinct timescales (e.g., microevolutionary versus macroevolutionary timescales). This spatial and temporal variability can introduce nonuniform selective pressure on the individuals who happen to experience it.

What has been said so far is not up for dispute, nor should it be. The point that Matthewson and Griffiths aim to emphasize by discussing such environments is that species have adapted in sometimes surprising ways to deal with environmental variability and uncertainty, not all of which manifest as comfortable or ideal states of being for their members:

> [The common monkey flower (*Mimulus guttatus*)] has evolved traits to deal with the variable circumstances that regularly arise in this environment. One such trait is early flowering. If the plant is growing in a poor enough location with respect to soil nutrients and water availability, its life span and growth potential will be limited, and the optimal age and size for reproduction is thereby reduced. *This is not a good situation for the plant to be in*. It will produce less seed than a conspecific that grows longer and larger before flowering. But for the less fortunate plant, that strategy would incur too great a risk of not reproducing at all. *The plant's developmental mechanisms are designed to make the most of a bad situation.* (Matthewson and Griffiths 2017, p. 454; my emphasis)

Spatial and temporal heterogeneities, although often glossed over in philosophical discourse, have proven the norm in nature. There are many

---

[20] This of course assumes (i) the absence of temporal autocorrelation and (ii) that there is no correlation between exhibiting particular character states and the probability of encountering specified environments.

examples showing how organisms adapt in ways that balance their lifetime prospects for reproduction against the risk of catastrophic failure (death or extinction). Some annual plants, for instance, have evolved "seed banks," which enable them to stagger or delay seed set and germination. This so-called "bet-hedging" strategy can be a highly advantageous in the event of unpredictable seasonal weather conditions that are severe enough to inhibit growth (Brown and Venable 1986). From an evolutionary perspective, having the ability to "make the best of a bad situation" often involves conservatively adopting a strategy that maximizes realized lifetime fitness (geometric mean fecundity) by reducing variance in reproductive output. Expected (arithmetic mean) reproductive contribution for any individual might well be lower than it otherwise could be for such a strategy. There are, after all, alternative strategies that could do better when conditions are favorable. However, the risk of altogether failing to reproduce is considerably reduced.

The outcome of selective evolution on such life history traits is a topic to which we will turn later in this section. Before moving on, however, it is well worth restating what this "third way of going wrong" contributes to the prevailing dialectic surrounding attributions of pathology. It enables us to determine on biological grounds alone whether things have gone better or worse for an organism in relation to its conspecifics and the state of its environment. Importantly, this determination can be made even when all conspecifics are performing their selected function (e.g., developing dwarfishness) in an environment that is historically normal. Mirroring the circumstances of nutritionally deprived, low birth weight human babies who develop a so-called "thrifty" compensatory phenotype, unfortunate (dwarf) monkey flowers are *worse off* than counterparts who are more likely to grow larger and produce greater numbers of seeds in good conditions. Yet, their misfortune is neither the result of mechanism(s) failure nor environmental mismatch as traditionally conceived. The developmental response to environmental stressors is the manifestation of a heritable, adaptively plastic strategy. Given inhospitable circumstances and their particular legacy of selective evolution, dwarf monkey flowers do indeed "make the best of a bad situation."

The fourth and final way of going wrong bears a strong resemblance to the preceding one. It, too, reveals a way that an organism can fail to optimally fit its habitat. Matthewson and Griffiths call the fourth way of going wrong "heuristic failure" (2017, p. 456). Unlike the previous way of going wrong, disorders due to heuristic failures occur in *hospitable* environments. Recall that the developmental response of the monkey flower in a normal but inhospitable environment is one that was already at hand or, rather, "in seed." Transgenerational selection has prepared monkey flowers for conditions in even the worst normal environments. They can accordingly afford to wait until a state of affairs is encountered before initiating the optimal developmental trajectory. When they act, their action is based on complete (or enough accurate) information about the state of the selective environment. In the case of disorders due to heuristic failure there is no such luxury. Some developmental "decisions" must either be made well in advance of anatomical separation (e.g., parturition) and often with partial or suspect information about environmental conditions. There are, for example, "developmental switches" for basic structural components of the body that must be "flipped" very early in the course of development during so-called "critical windows" if the ensuing cascades of complex interactions required for constructing an adaptive phenotype are to proceed properly. The alternative states of a switch can culminate in dramatically different phenotypes. Perhaps the classic example is the helmet and (tail-)spike morph of the water flea (*Daphnia cucullata*).[21] This "armor" enhances the chances of survival and, hence, reproduction in environments filled with predators. Yet, it is physiologically costly to produce and maintain. When there are few or no predators in the environment, water fleas have higher fitness without it. Even if there was available genetic variation for specialized variants (i.e., armored versus unarmored), the transgenerational evolutionary response produced by selection would almost certainly fail to keep pace with short-term ecological fluctuations in predator density. Selection has accordingly fashioned a plastic trait that responds to the presence of chemicals released by predators. However, the developmental response is triggered by chemical stimuli presented in the prior (parental) generation. Offspring exhibit the armored phenotype only if their mothers experienced an environment that was full of predators. As one might imagine, this arrangement does not bode well for water fleas born amid a transition from high to low numbers of predators. They incur a cost to their fitness even though they inhabit a safer (low predator density) environment.

What goes wrong here is that the evolved decision procedure (developmental rule) for water fleas confronts an unavoidable compromise. Other than by experiencing the environment into which they will be born, offspring cannot acquire the information required to develop an optimal

---

[21] The organizational and activational effects of hormones noted by developmental biologists are also "developmental switches" in this sense. Sex steroids in utero, for instance, can prepare organisms for the actions of their hormones later in development.

phenotype. But the time required to develop the adult phenotype that would confer a selective advantage prohibits such a waiting game. This optimization problem has been solved by selection for a rule of compromise in which the average fitness cost of a "false negative" (mistakenly predicting low predator density, so failing to produce armor) is higher than the average fitness cost of a "false positive" (mistakenly predicting high predator density, so producing armor).[22]

This type of error management or risk assessment is not unfamiliar to practitioners of Evolutionary Medicine, who refer to it colloquially as "the smoke detector principle" (Haselton and Buss 2000; Nesse 2001). Smoke detectors are notoriously oversensitive to the presence of smoke. We tolerate this hypersensitive bias because the cost of "under-sensitivity" (hyposensitivity) in the presence of an actual fire is often catastrophic. It is a cost that dwarfs the inconvenience associated with the response. When the cost of expressing an all-or-nothing response is relatively low compared to the potential harm it protects against, the optimal system will elect to express many false alarms. It is one of the central insights of Evolutionary Medicine that a predictable rate of error can accompany even those traits that have been selected for maximizing expected lifetime fitness. The cost of this imprecision is incurred by the individuals whose realized fitness goes unmaximized even in what is a very hospitable (e.g., low predator density) environment.[23]

With this final way of going wrong, Matthewson and Griffiths complete a set of four independently sufficient conditions for biological dysfunction. Note that the criteria are not intended as mutually exclusive; a disorder may well result from a combination of them. The proposed framework is much more refined than any Naturalistic or even hybrid precursors. This should come as a relief to critics of the dysfunction condition as originally expressed in Wakefield's work, especially for those who worry about the possible infiltration of rampant adaptationism into traditional medicine. Evolutionary Medicine, with its foundational appreciation for and expansion of Tinbergen's explanatory framework, is anything but naïvely adaptationist.

---

[22] The optimality model involved would obviously be much more complicated than suggested here. It would minimally involve the fitness cost of false positives, the fitness cost of false negatives, the relative fitness benefits of accurate prediction or "matching," as well as probabilities of offspring environments matching parental environments, among other considerations.

[23] Bourrat and Griffiths (2021) have recently suggested that the three ways of going wrong other than mechanism failure might be characterized as "mismatches" on distinct temporal scales (physiological, developmental, and evolutionary).

4.4 Brothers in Arms: The Organizational Account of Function

The formidable philosophical edifice erected by Matthewson and Griffiths can be further bolstered by the Organizational Account of function (Montévil and Mossio 2015; Saborido and Moreno 2015; Mossio and Bich 2017). The Organizational Account complements their view by providing a detailed characterization of the mutual interdependence that we observe among an organism's constitutive parts. Organisms *qua* biological entities actively maintain their own existence and integrity by increasing entropy in their surrounding environments. This can only be accomplished if organisms are thermodynamically open systems. Energy and matter must permeate their boundaries as well as be transformable and transportable within them. The flow of energy and matter cannot go uncontrolled, however. An organism's constitutive parts, along with their overall structural arrangement, constrain the process along varying but specific timescales. The set of constraints embodied by the organism has two crucial features. The first is that any constraint must channel the flow of energy and matter in such a way as to maintain the conditions of existence for other constraints. Using the terminology of the Organization Account, it is thereby "generative." Proponents of the view argue that a constraint must also be "dependent," such that the existence of a constraint necessarily depends on the action of other constraints. Systems, as in sets of constraints, that are both generative and dependent supposedly exemplify a (non-vicious) form of explanatory circularity known as "organizational (or causal) closure." According to the Organizational Account, organisms maintain a relatively stable existence (persist) because they realize "closure of (causal) constraints."

Despite some rather daunting differences of terminology, the Organizational Account of function is largely consistent with the Evolutionary Medicine framework. A structure is dysfunctional on the Organizational Account when (i) it fails to channel the flow of energy and matter in a way that would maintain the conditions of existence for other constraining structures or (ii) its persistence is unaffected by changes in the number, membership, or magnitude of constraints that collectively bound the organism. Put succinctly, dysfunction must be a consequence of structural "degeneracy" or "independence." The term "disorder" is reserved for dysfunctions that compromise the persistence or relative stability *of the individual organism*. It emphasizes maintenance over reproduction.

At face value, this seems at odds with the selective explanation of dysfunction already on offer, which often stresses the impact on survival and reproductive capability. There is no genuine conflict, however. Life history theory (Stearns 1992) is the foundational theoretical framework for Evolutionary Medicine and its philosophical exponents. As noted in Section 2, the fundamental postulate of life history theory is that organisms are faced with a finite amount of available energy (resources) that must be competitively allocated to the functions of maintenance, growth, and reproduction. This allocation procedure requires unavoidable compromises. Energy allocation to one function necessarily deprives the remaining two demands. By way of example, energy directed toward reproduction (e.g., energy expenditure on mating effort) is no longer available for investment in maintenance (e.g., homeostatic regulation) or growth (e.g., cell proliferation or development). These fundamental constraints are referred to as "life history trade-offs." For each given stage of the organismal life-course in any environment, natural selection will favor the energy allocation strategies that maximize lifetime fitness. While the strategies are usually presented in the form of mathematical optimization functions, these abstract constructs are realized by so-called "life history traits (or characteristics)." The number of life history traits and the variance associated with them depend on the species of interest. Among the life history traits studied in humans are adult body size, fertility rates, length of lifespan, senescence, time of gestation, age of weaning, juvenile mortality profiles, age at (sexual) maturation, and onset of menopause (as mentioned in Section 2). The mechanistic constraints that govern trade-offs among such phenotypic traits may be physiological, developmental, or genetic.

This theoretical framework requires consideration of fundamental developmental (growth) and physiological (maintenance) functions. These two basic functions together account for organismal persistence, which is also the defining feature of the Organizational Account. The philosophical analysis of (dys)function proposed by the Organizational Account identifies possible alternative strategies of energy allocation to maintenance, or what that account would call "regimes of causal closure."[24] These can be extended to explore the subsequent evolutionary dynamics associated with promoting the viability (contra fecundity) component of organismal fitness. Function ascriptions, on the Organizational Account, thus represent a shift from Tinbergen's ultimate, backward-looking

---

[24] It is telling that regimes of causal closure are sometimes referred to as "regimes of self-maintenance" (Mossio, Saborido, and Moreno 2009, p. 829).

questions of "survival value" to his proximate, forward-looking questions of "causation." As such, the Organizational Account has much in common with the Causal Role account of function for a component embedded in a system (organism) with the capacity to survive and reproduce.

Recall, again, that Matthewson and Griffiths' approach allowed for disorder even in cases where there is no apparent *evolutionary* dysfunction. Disorder of trait or state can also emerge when one restricts consideration solely to the *prospective fitness* of token individuals that currently exhibit suboptimal trait types. These individuals may in fact be "worse off" in the sense that they are less likely to survive or successfully reproduce. While natural selection cannot improve their lot in life, modern medicine can attempt to ameliorate. But whether and how we choose to intervene is predicated on the biological judgment that the individual's prospects for survival (maintenance) are reduced. The Organizational Account of function helps justify this by drawing our attention to the physical (thermodynamic) and biochemical (physiological) limits of orderly functioning for the members of a species. Its characterization of dysfunction as an intrinsic condition that reduces healthy longevity is thus strikingly consistent with the "longevity first" view adopted by traditional medicine. Individual human flourishing may well turn a blind eye to the reproductive imperative, but its very possibility presupposes survival.

It should be evident that proponents of early hybrid accounts (e.g., Wakefield) can embrace the proposed revisions to the evolutionary dysfunction condition suggested by Griffiths and Matthewson's account and the Organizational Account. In order to do so, however, they must relinquish the strong requirement that there be disorder only when there is mismatch (in the traditional sense) or mechanism failure. From a more sophisticated evolutionary perspective, there is nothing "dysfunctional" about the imperfect responses that are generated by normal but inhospitable environments or heuristic failures in hospitable environments. Such responses are the result of predictable allocation strategies. These strategies are considered distinct "regimes of causal closure" for the Organizational Account. The evolutionary medicine community and its chief philosophical exponents are cognizant of the pitfalls associated with naïve adaptationist thinking. They recognize that some disorders and susceptibilities may not be directly adaptive for the individuals that exhibit them. A disordered state or susceptibility may well be the by-product of selection on another function or mechanism. Nesse (2019) has recently coined a name for the crude adaptationist mistake of overlooking such possibilities: "Viewing Diseases as Adaptations." A broader and deeper understanding of a (disordered) trait and the system in which it is embedded, along the lines made possible by Tinbergen's four questions, is necessary in order to understand how an evolutionarily non-dysfunctional trait can still leave us compromised.

4.5 Are There Grounds for Reconciliation with Normativism?

This biologically rejuvenated hybrid account, which I call the "Life History-Organizational Account" (LH-OA) of disorder, renews the possibility of reconciliation with those who would otherwise opt for some form of Normativism. A short detour through philosophical methodology helps show the way.

Perhaps the greatest obstacle to progress in the philosophy of medicine has been the meta-philosophical commitment to conceptual analysis. As several philosophers (Schwartz 2007; Murphy 2006; Lemoine 2013) have argued, that approach was doomed to failure from the outset. Effectively implementing it would require the existence of an agreed upon extension class of diseases, medical disorders, or pathologies. The success of any conceptual analysis is determined by whether or how well the class of disorders defined by its proposed set of individually necessary and jointly sufficient conditions matches this consensus class. If an analysis fails to include some agreed upon disorder, the set of definitional conditions it proposes are deemed too stringent or "strong." Alternatively, conditions of an analysis are too permissive or "weak" when they define physical states that fall outside the agreed upon class as being nonetheless disordered. The most pressing difficulty for this general approach is the discrepancy between the class of physical states that laypersons might consider disordered and those that biomedical professionals would consider pathological. In many ways, it is this disparity that has set the course of philosophical debate about health and disease. Those committed to Normativism point to historical instances showing how erroneous or bias-infused science can lead to unwelcome outcomes (e.g., homosexuality as "pathological") and consequently plead for a corrective return to a more general value-laden conception of health and well-being. Advocates for Naturalism have responded by refining their analyses to include only purportedly *unambiguous* cases of disorder and exclude supposedly *clear* instances of health. The trouble is that certainty regarding the medical status of particular cases is sometimes determined by the very definitional conditions whose adequacy is in question. This results in undecided or controversial cases. What a Naturalistic account considers to be a clear case of disorder might be an unambiguous instance of normal or healthy functioning on a Normativist account. Without a preestablished consensus class of disorders and a class consisting of healthy states to complement it, the only foreseeable outcome of traditional conceptual analysis in philosophy of medicine is a paralytic stalemate.

One seemingly sensible way to resolve this impasse is to engage in explication rather than conceptual analysis (Carnap 1950). Explicating the concept of disease, for instance, restricts focus to the systematic usage of this concept (or cognates) in professional biomedicine. Doing so would emphasize the many and diverse theoretical roles of the concept as revealed by the terms that practitioners use to pick it out in various contexts. This seems like a welcome restriction insofar as it is largely to the expertise of medical professionals that we turn when seeking remedy for physical disease. However, as some Normativists might see matters, this proposed turn to explication imposes precisely what they seek to resist: the hegemony of scientific expertise in medicine. They have gone to great lengths to show how it can marginalize or even undermine the interests of individuals and minority groups.

It is unarguable that broad social norms and non-epistemic values have *occasionally* misled science. Historical instances of this tend to weigh heavily on the thinking of those who favor some form of Normativism. Enumeration of cases only serves to reinforce their preferred inference, despite a general lack of knowledge about the relative prevalence (base rate) of instances where science has supposedly been so sullied. What fundamentally seems to motivate the Normativist's resistance (to the project of explication) is the worry that *any* infusion of non-epistemic norms into biomedicine threatens to undercut the precedence of self-determination in judgments of health. Undercutting this is apparently something that Normativists cannot abide. Health is supposed to be a deeply personal matter. The individual, at least from a Normativist perspective, is just the smallest possible minority. A minority group's say about health should not be so easily overturned by any majority group, scientific or otherwise. Upholding individual authority and self-interest is, for many Normativist accounts, seemingly the best prospect we have for guarding against the tyrannical values of an ideological majority dressed in the garb of science.

If the foregoing is correct, the refinements to the traditional dysfunction condition proposed by the LH-OA framework should provide a modicum of relief for Normativists. We now have more specific, principled constraints on membership in the class of medical disorders. Any condition, if indeed it is to be a genuine disorder, must have an objective biological basis in at least one of the four "ways of going wrong": (i) mechanism failure; (ii) environmental mismatch; (iii) normal but inhospitable environments; and (iv) heuristic failure. A condition whose undesirability cannot be explained by reference to at least one of these is "unhealthy" in only a figurative or merely stipulative sense. We accordingly have an expanded repertoire of principled (biological) reasons for medically disvaluing some conditions. On the one hand, this enables us to potentially count as diseased those conditions that were once considered unviable candidates because their undesirability had no obvious physiological or evolutionary basis in mechanism failure or traditional mismatch. Individuals consequently have more conceptual resources to draw on when attempting to justify the claim that they are unhealthy. On the other hand, societies must now meet more stringent evidentiary standards by specifying which of the biological "ways of going wrong" underwrite medical determinations of disease. The fact that our society disvalues a condition must be more than a matter of the mere fact that we happen to dislike it or consider it deviant. Refining the bioscience helps safeguard against the baseless attribution of disease.

Deciding exactly how much weight to give the self-reports of individuals or minorities when it comes to determining health status remains a vexing problem. Ultimately, it is a matter that can only be resolved by addressing the evaluative component (i.e., harm condition) of a hybrid theory. This difficulty will be addressed in the concluding section. The takeaway from the current section, though, is that the dysfunction condition for a hybrid account of disorder can in fact be defended in the face of criticism that habitually targets more dated methods of evolutionary analysis. Worries about the dysfunction condition that may once have driven some critics into the open arms of Normativism are much less compelling than has sometimes been suggested.

5 From Sensible Normativism to a Hybrid Account

5.1 From Nature to Normativity

The overarching aim of this Element is to articulate something that resembles the state of the art when it comes to hybrid accounts of disorder. I have shown that the biological sciences, especially evolutionary theory, have advanced well beyond what early versions of the Bio-Statistical Theory, Causal Role, or even Selected Effects accounts assumed. Arguably the most important insights have come via research in Evolutionary Medicine. The theoretical foundation for this burgeoning area of biomedical inquiry lies in evolutionary theory, primarily life history theory. If nothing else, then, this Element suggests that LH-OA should serve as the appropriate target for the ire of philosophical critics.

A lingering concern nonetheless remains for those who believe that the criteria for disease and disorder suggested by a hybrid account represent the best hope for progress. Barrowing Wakefield's terminology, the worry is one over the so-called "harm condition." This patently normative component has been mentioned only in passing. The silence is deafening. A hybrid theory's refusal to engage seriously with this vexing issue would be no less frustrating than the Normativist's use of the Bio-Statistical Theory or early Selected Effects accounts as a foil. Difficult as the issue may be, it is high time for hybrid theorists to address the issue explicitly. How and to what extent do values play a role in determinations of disease?

The concluding section of this Element has two goals. The first is to illustrate exactly how the philosophical LH-OA framework can do justice to what actually transpires in biomedicine. This will be done with the aid of a much more detailed example than those presented earlier. On the one hand, this framework has a direct theoretical bearing on the diagnosis and nosology (classification) of disease states. On the other, it transforms the way we treat disorders, with corresponding changes of emphasis on preventative (harm reduction policies) and reactive interventions. The progress wrought by the LH-OA framework only becomes evident when closely examining these two stages of biomedical practice.

The second goal is to make the role of norms in a hybrid account explicit. This will require distinguishing two different strands of Normativism (SNC), Internalist versus Externalist, and showing that only one of these (Externalist SNC) is in fact a reasonable alternative to hybrid accounts. The hybrid account I favor can profitably endorse the role but not the precedence that that form of Normativism envisages for broad social values. However, even this remaining form of Normativism has seemingly unacceptable consequences that the LH-OA framework can apparently avoid.

5.2 LH-OA in Action: A Detailed Example

Before diving into more perilous normative waters, let us focus on the conceptual improvements that are ushered in with LH-OA. A particularly enlightening example concerns improvements to our understanding of human adaptations to living at high altitudes and how this knowledge influences the determination of medical thresholds for anemia (Beall 2019).

It is estimated that 81.6 million people, 1.07% of the global population, live at altitudes at or above 2,500 meters (Tremblay and Ainslie 2021). Survival at higher altitudes is notoriously difficult because atmospheric (barometric) pressure is lower than it is at sea level, which consequently reduces the concentration and thus availability of oxygen molecules in every breath we take. For example, the average concentration of oxygen per breath at an elevation of approximately 5,000 meters is reduced to just fifty percent of what it would be at sea level. As there are absolute mitochondrial requirements on the steady delivery of oxygen to tissues, reduced oxygen saturation of the blood and in the brain places severe stress on the metabolic functions of all our cells. It comes as no surprise, then, that many lowland natives experience debilitating symptoms of what is often called "Acute Mountain Sickness" when ascending to elevation rapidly and residing there for prolonged periods.

Vertebrates show a variety of adaptations that enable them to meet their physiological requirements for oxygen in such circumstances (Bouverot 2012). There is also considerable variation among human populations, which show distinct molecular, genetic, and physiological hemoglobin responses at high altitudes. The contrast between native dwellers of the Andes Mountain Range of South America and native populations of Tibetan/Nepalese in the Himalayas has proven particularly illuminating. Andean highlanders have lived at altitude for approximately 11,000 years (Petousi and Robbins 2014). This span provides ample time for selective evolution in human populations. Studies show that these Andeans have adapted to high-altitude hypoxia by dramatically increasing the production of hemoglobin. This response is consistent with findings for the response of native (non-Andean) lowland populations who have more recently (within the last 500 years or so) migrated to higher altitudes. They, too, show increased concentrations of hemoglobin, although not to the same extent nor over as great a range of heights as observed for Andean highlanders. Together, these findings seemed to lend support to the selective hypothesis that there is a single optimal or universal response to high-altitude hypoxia in humans: increased synthesis of hemoglobin.

The assumption that there is selection for a single optimal response has proven hasty, however. It had all the trappings of a naïvely adaptationist "just so" story of the kind that researchers have been routinely warned against. Some human populations can survive and even thrive at high altitude despite showing no significant increases in hemoglobin and red blood cell concentration. Most notable among these are Tibetan highlanders, including the Sherpas of Nepal. Populations have resided at elevations in excess of 3,500 meters for more than 25,000 years (Petousi and Robbins 2014). Although they can exhibit marked increases in the production of hemoglobin and red blood cells at extremely high altitudes (usually well in excess of 4,500 meters), this is not the observed adaptive response to low levels of oxygen across the range of altitudes (3,500–4,500 meters) historically commonplace for them. These highlanders instead compensate with increased resting ventilation (i.e., hastening the exchange of air between the lungs and the ambient air), less hypoxic pulmonary vasoconstriction (which diverts blood to better ventilated areas), and higher levels of nitric oxide in exhaled breath and blood, which correlates with lower pulmonary artery pressure (Beall 2014). In effect, these physiological measures increase cardiac output and, thereby, hasten the acquisition and effective transport of oxygen.

Let us now turn our attention to anemia. Anemia is defined as a pathological condition due to too few red blood cells or too little hemoglobin for adequate oxygen transport. It is unarguably pathological insofar as it contributes to poor growth and development as well as lowered exercise/work capacity. While this pathology may be due to low oxygen conditions at high altitude, it can also be caused by insufficient iron. Iron is an indispensable elemental precursor for the biochemical synthesis of hemoglobin and red blood cells. The human body makes less hemoglobin and fewer red blood cells when faced with lower levels of iron, which can obviously stress oxygen delivery. Approximately a quarter of the global population suffers from iron-deficiency anemia.

The 2018 World Health Organization guidelines for the diagnosis of anemia incorporate the findings for native Andean highlanders and transplanted (non-Andean) lowlanders noted earlier (World Health Organization 2017). Based on those findings, the WHO recommended a single table of standard adjustments for altitude. The basic presumption underlying this standardization was that there must be a gradual increase in average hemoglobin/red blood cell concentration with increased altitude, the degree of which varies with respect to the amount of time that a population has resided at altitude. The problem with this expectation and, hence, the suggested corrections is that the findings for Tibetan/Nepalese highlanders do not conform to it *even after controlling for the possibility of iron-deficiency*. Applying the suggested WHO corrections for altitude would make it such that 77.8% of Tibetan men and 86.5% of Tibetan women are apparently anemic despite no indication of iron deficiency or other confounding factors such as chronic inflammation (Sarna, Brittenham, and Beall 2020). These Tibetans obviously do not suffer from anemia. Other factors notwithstanding, they are healthy. The road to misdiagnosis has, it seems, been paved with good intentions.

This miscalculation can have serious consequences in cases where WHO guidelines inform the treatment of those deemed anemic. It has recently been shown that supplementing the diet of Tibetans/Nepalese highlanders with additional iron to increase hemoglobin concentration may in fact yield detrimental effects. In Tibetan men, for instance, increasing hemoglobin via iron supplementation reduces exercise capacity by lowering oxygen delivery (Wagner et al. 2015). Furthermore, Tibetan women have worse pregnancy outcomes (i.e., fewer live births and more preterm and small-for-gestational-age births) when their hemoglobin concentrations exceed the population average (Cho et al. 2017). Such outcomes diminish components of fitness (survival in men, fecundity for women) and implicate the influence of selection against genetic variants that increase the production of hemoglobin.

Genetic and molecular evidence strongly supports this selective hypothesis. There is heritable genetic variation (single nucleotide polymorphisms) in several genes that interact along the oxygen homeostasis pathway, which intersects with the iron homeostasis pathway. The responses to oxygen deprivation resulting from high-altitude hypoxia and iron-deficiency anemia happen to share these molecular pathways. Among the more pertinent genes are *EGLN1* and *EPAS1*. All of our cells have oxygen-sensing proteins called prolyl hydroxylase domain proteins (PHDs). One of these, PHD-2, is encoded by *EGLN1*. When there is an adequate supply of oxygen, PHD-2 will prevent the accumulation of a constitutively expressed transcription factor known as Hypoxia Inducible Factor-2 (HIF-2) by marking it for degradation. As the name suggests, HIFs play a crucial role in the way that cells adapt physiologically to hypoxic and nutrient-limited conditions by upregulating transcription. HIF-2 also features in the iron homeostasis pathway. Accumulation of HIF-2 can induce the transcription of hundreds of genes. Notably, it induces the transcription of *HAMP* (hepcidin), which allows for increased intestinal absorption of iron, and the gene *EPO* that encodes the hematopoietic growth factor erythropoietin for regulating the growth and maturation of red blood cells. HIF-2 has two protein subunits, one of which (HIF-2α) is variable. This variable subunit exhibits known genetic variants or alternative alleles at the *EPAS1* locus. Tibetan/Nepalese highlanders have unexpectedly high frequencies of unique genetic variants at *EPAS1* and *EGLN1*, as well as uncommonly high allele frequencies at some of the target genes for HIF-2 that are involved in the synthesis of hemoglobin. A unique mutational variant, often referred to as the "major Tibetan" allele, of the *EGLN1* gene exhibits heightened sensitivity to oxygen molecules and subsequently maintains normal levels of PHD-2 degradation (of HIF-2) even as oxygen becomes less readily available. Reducing HIF-2 levels curtails erythropoietin synthesis, which consequently scales back the overall concentration of hemoglobin (Gassmann et al. 2019). In other replicated studies, it has been shown that Tibetans who are homozygous for the high frequency Tibetan-specific variant of *EPAS1* have *un*elevated hemoglobin concentrations that are on average nearly 12% (~2.0 g/dL) lower than those found in fellow population members who are homozygous for the alternative ("minor") ancestral allele (Beall et al.2010; Beall 2014).

There obviously remains much scientific work to be done when it comes to fully understanding the mechanisms that underpin oxygen homeostasis and corresponding attributions of anemia. I have restricted discussion here to findings in just two of the highland regions that are intensively studied. How, then, does the LH-OH framework help us understand the interplay between biology and medical theory in this case?

First, the LH-OA account of dysfunction begins to make sense of anemia as a pathological dysfunction and, therefore, medical disorder by identifying an objective threshold for healthy functioning in the individual's basal metabolism. Individuals have minimal mitochondrial oxygen requirements. This requirement may of course vary among individuals within a population. It will also vary temporally across a person's life. However, it must be considered a minimal physiological threshold for the survival of an individual at any specified point in time. Falling below this threshold compromises an individual's viability. That would, following the Organizational Account of function, be a failure to channel the flow of energy and matter in a way that maintains the conditions of existence for other constraining structures that together ensure the persistence of the individual as a bounded system characterized by the "closure of (causal) constraints." Anemia is a pathological condition *for the individual*, a medical disorder if ever there was one.

The second way that LH-OA account better captures the interplay between biology and medical theory involves its reliance on evolutionary theorizing. As explained in the previous section, the "Life History" component of the LH-OA account introduces expanded evolutionary considerations into determinations of dysfunction. Proposing the use of evolutionary theory *tout court* is, of course, nothing new. Various Selected Effects accounts would already suggest as much. However, an application of evolutionary theorizing that expressly encompasses ecological and developmental processes acting on different spatial or temporal scales is a more recent innovation. This is an all important advance because the Organizational Account alone, with its emphasis on physiological and homeostatic regulation (i.e., Tinbergen's question of "causation"), does not speak to the issue of functional (as in "non-dysfunctional") variation among individuals, populations, or species. Focused as Organizational Account is on the integrity and persistence of individual organisms, it would have little to tell us about the following sort of question: Why do populations of Tibetan/Nepalese highlanders respond to reduced oxygen availability by doing X instead of Y (as, say, Andean highlanders do) when doing Y apparently suffices?

Explanations of such non-dysfunctional variation emanate readily from evolutionary theory. This expansive body of theory can explain why we see the available alternative responses that we do. The most obvious consequence of this frameshift is the more nuanced array of ways that biological components/systems can be classified as functional or dysfunctional. In the case of high-altitude hypoxia, for instance, evolutionary theorizing sheds light on why it is appropriate to characterize elevated hemoglobin concentrations in Tibetan/Nepalese highlanders as dysfunctional (i.e., fitness reducing) even though that physiological response is appropriate for most Andean highlanders. Furthermore, it raises fruitful questions for further inquiry. One such question concerns the biomechanical basis of the observed fitness cost of elevated hemoglobin concentration *in Tibetans/Nepalese*. A promising line of inquiry suggests that the poor pregnancy outcomes and reduced exercise/work capacity that come with increasing hemoglobin may be due to the toll that pumping more viscous, red blood cell-rich blood takes on the ventricles of the heart that push it. Tellingly, this insight has been gained by examining the variance *in Andean highlanders*, who tend to show diminished fitness when hemoglobin concentrations exceed the population average. Identifying heritable variation within or between populations is a prerequisite for selective explanation. Insights from genetics and anthropology are indispensable for identification. Modern genetic techniques (e.g., candidate gene analyses, genome-wide and epigenome-wide association studies, and molecular sequencing) have revealed important clues about the historical origins of distinct populations, such as which allelic variants were ancestral precursors. These clues provide crucial information about the strength of natural selection and other evolutionary factors as well as the durations over which these have operated. Not only does work in evolutionary anthropology often corroborate these findings; it also chronicles the ways that collective human activity as a form of "ecological niche construction" shapes adaptation. In so doing, this work helpfully reveals factors that may potentially confound overly simplistic selective explanations.

Third, the LH-OA account of dysfunction makes sense of the changes to diagnostic criteria for anemia. General recognition of adaptation to high-altitude hypoxia was a welcome step forward. But initial failures to entertain the possibility of distinct human adaptations to the same stressor (low oxygen availability) resulted in a profound overestimation of the prevalence of anemia. Experts from a broad range biomedical disciplines are currently in the process of remedying this and issuing revamped guidelines for the diagnosis of anemia. Despite this, there are no clarion calls for the elimination of anemia as a genuinely pathological condition. Anemia is still defined as a pathological condition caused by having too few red blood cells or too little hemoglobin to meet physiological (mitochondrial) oxygen requirements. The criteria for its determination have been refined by taking into consideration unique genetic and regulatory profiles for the iron-homeostasis and oxygen-homeostasis pathways as well as the ecological and cultural differences among populations that may have driven divergent selection. The class of individuals who are considered genuinely anemic consequently includes fewer members. No one has or should complain about this achievement. It is all in the course of standard scientific practice. And it has been accomplished on objective scientific grounds, without the questionable imposition of non-epistemic cultural norms.

Fourth, the LH-OA account of dysfunction justifies what many consider to be "best medical practice" in the going case. Most Tibetan/Nepalese highlanders are not, as was once thought, anemic. The lower hemoglobin and red blood cell concentrations they exhibit are not dysfunctional. Physico-chemically established thresholds for organismal integrity (viability) and evolutionary considerations (fecundity) together help explain why this is so. The prognosis for Tibetan/Nepalese highlanders has shifted accordingly. Harm reduction policies have followed suit. Iron supplementation via diet or blood transfusion is no longer a suggested corrective intervention for Tibetan/Nepalese highlanders. It is now widely recognized that such treatment is potentially harmful for them. Scarce resources once bound for investment in the unnecessary *treatment* of apparent anemia can instead be redirected toward *preventing* other legitimate threats to health in these communities. There is, for example, considerable evidence suggesting that the prevalence of noncommunicable diseases (e.g., diabetes mellitus, obesity, hypertension) and risk factors for the development of such diseases at high altitude differ from lowland areas and are steadily increasing (Koirala et al. 2018). Tibetans, in particular, appear to have a high prevalence of dysglycemia, which increases the risk of developing diabetes and cardiovascular disease (Okumiya et al. 2016). This is but one obvious target for future research.

5.3 Is LH-OA Preferable? From Theory to Practice and Back Again

The advances that come with the LH-OA account of (dys)function are legion. While I cannot detail its advantages in all or even most cases here, I believe that the conceptual framework for biological *dysfunction* it proposes currently has no equal. Demonstrating its sufficiency as an account of *disorder* is a more difficult matter. Of paramount importance for accomplishing that aim is addressing the issue of value-ladenness. That requires disambiguating two forms of Normativism and examining the distinct roles that norms play in each. I will argue that LH-OA's accommodation of norms ultimately requires fewer additional justificatory claims and thereby seems to better accord with the basic intuitions of medical professionals and possibly laypersons alike.

The first version of Normativism is by now familiar for readers, as it is the commonplace version of the approach (SNC) introduced in Section 1. It typically vets the historical record for episodes demonstrating how non-epistemic values influence the classification of conditions as diseases. The medicalization of homosexuality is a particularly striking example, but various disabilities (e.g., deafness) and cognitive disorders (e.g., schizophrenia) also feature prominently in the literature. I will refer to this version of Normativism as "Internalist" on the grounds that the influence of social norms and (non-epistemic) values on disease determination is most directly asserted from within the medical sciences.

The primary shortcoming of Internalist SNC is that it fails to provide principled reasons for adopting the episodic (historical) boundaries that it demands. It overlooks or simply ignores the advances that can accompany scientific enquiry over longer periods. At one time or another, all phenomena of interest to science or proto-science are ill-defined. A large part of science involves working out how best to make discrete (categorical) a world that typically presents itself to us as a nexus of continua. This process of understanding takes time. A LH-OA hybrid theorist can reinterpret what Internalist SNC takes to be a clear instance of scientific standards establishing apparent disease as a case in which subsequently revamped scientific standards prevail in spite of dominant social norms. Without further argument, it is not obvious why philosophical conclusions based on the "march of science" should be held captive by, what proponents of Internalist SNC consider, particular episodes of historical intrigue. Furthermore, even if we accept the demarcation of episodes that Internalist SNC has thus far proposed, it remains unclear whether such cases will prove to be the rule rather than the exception. Internalist SNC would have us believe that social norms and values influence disease determination in most if not all cases. While a more traditional Naturalistic (ONR) account of disease might have been undone by just a single case, this inductive inference must be sustained in order to undermine the plausibility of hybrid accounts that also embrace value considerations. To demonstrate this convincingly, however, Internalist SNC would have to establish a comprehensive set of carefully delimited cases of disease determination and show that their favored characterization holds in most of these cases. At best, Normativists of this ilk usually generalize on but a small number of (sometimes) carefully researched cases, most of which rely on scientific theories or practices that have long since been superseded.

The use to which cases like homosexuality are put raises yet another worry about Internalist forms of Normativism. Such cases have typically been presented as "cautionary tales" that supposedly undermine Naturalistic accounts of disease. What, then, would make these fit for purpose? The answer must be that these scientific judgments of pathology failed to portray the situation *as it actually was(is)*. Most would agree that homosexuality was never appropriately conceived of as a disorder. However, this seemingly sensible outlook creates surprising tension for Internalist SNC. Its proponents want to depict medical science as nothing more than a particularly effective tool for the social construction of reality. By its own tenets, then, medical science supposedly has no additional or special authority beyond that licensed by prevailing social norms. The scientific standards and practices unique to medicine are portrayed as independently reinforcing, justifying, or legitimating social norms. The crux of these cautionary tales is, after all, that medicine purportedly "discovers" or "confirms" that these prevailing norms were correct all along. However, if social norms or values ultimately determine "matters of (medical) fact," there is no sense to be made of the claim that, for example, homosexuality was *mis*diagnosed as a disorder. Unless, of course, one falls back on the belief that biomedical science can somehow correct our socially licensed ontology. But, again, biomedical science simply doesn't have that sort of independence or authority according to Internalist SNC. This particular version of Normativism consequently trades on the epistemic authority that science presumably has when it comes to accessing the world as it actually is. If medical science is merely at the disposal of society at large when it comes the construction of pathology, then homosexuality was in fact a "disease" in the past but not the present. Pathology problematically hinges on the whims of society at large.

The foregoing "Internalist" version of Normativism should be juxtaposed with an "Externalist" counterpart. Arguably the most sophisticated version of the latter approach is to be found in the work of Shane Glackin and collaborators (Glackin 2010, 2019; Conley and Glackin 2021). Whereas Internalist Normativism attempts to show that social norms misappropriate biomedical theory and practice, an Externalist version recognizes that it is often not obvious whether, how, or with what frequency this happens. Externalists offer a more rudimentary argument for Normativism whereon social norms and values are superimposed upon what is presumed to be a pristine biomedical edifice. An Externalist account of disease determination can accordingly be maintained even if it is granted that the workings of biomedicine are mostly unencumbered by the normative machinations of society at large. While perhaps compatible with Internalist versions of SNC, what importantly distinguishes Externalist SNC is the claim that the unique role of social norms and values in disease determination is a direct consequence of their *nonmedical* status.

How exactly do such "external" (nonmedical) factors influence disease determination? Glackin (2019) and Conley and Glackin (2021) provide considerable clarity on this issue. Medicine, for them, is not unlike engineering in the sense that it aims to shape the world according to the way we think it should be. The aims of medicine are correspondingly much more *prescriptive* than *descriptive*. The desires of the subject or those of a collective reached by means of rational intersubjective discourse (in a democratic society) are supposedly what give direction to this change (Glackin 2010).

Four normative conditions feature indispensably in disease determination (Conley and Glackin 2021, p. 10) on this view. A biological/behavioral state is a disease if and only if it is regarded,

(1) as presenting an *intolerable* state of affairs;
(2) as not representing a *moral failing* of the individual;
(3) as *not being worth* reorganizing society so as to fully neutralize the relative impairment that accompanies it; and
(4) as *being worthwhile* to divert resources to "correct" and/or ameliorate it.

The italicized terms reveal unambiguously normative considerations or moral judgments of the kind that pertain to social institutions generally rather than any of the sciences in particular. This set of criteria constitutes the procedure for "social classification" that must be superimposed upon biomedical classification in order for a biological state to be considered a genuine disease. Modern biomedical science is perfectly well suited for answering what Glackin (2019, p. 260) calls "constitution questions," which concern the physical basis of a patient's condition. The answers typically take the form of lists that include a wide range of empirically verifiable diagnostic criteria. Externalist SNC acknowledges that these are the objective (mind-independent), natural phenomena that ground judgments of disease. These are the factors that traditional Naturalistic accounts of disorder/disease would emphasize as being sufficient. However, for Externalist SNC, answers to such constitution questions importantly fail to address general "status questions." These questions concern "what must be true of an individual if s/he is to be reasonably attributed the status of having a particular disease" (Glackin 2019, p. 260). It is allegedly a category mistake to posit physical facts about a patient's condition as an answer to this latter sort of status question. Having a disease apparently involves something "over and above" exhibiting the physical (diagnostic) criteria that are deemed necessary. This additional something, for Externalist SNC, is the "anchoring fact" that society judges or regards the physical conditions which ground a disease as having satisfied the four conditions noted above. In other words, society has taken a (normative) stance toward the set of diagnostic criteria. The physical (diagnostic) facts are accordingly taken to ground a disease because society has evaluated them as such. This *procedural fact* regarding the observance of rules for rational discourse makes all the difference for Externalist SNC. Virchow's infamous quote seems to loom large once again: "Medicine is a social science and politics is nothing else but medicine on a large scale" (Taylor and Rieger 1985 [1848]).

What makes Externalist SNC more imposing than its Internalist counterpart is that it can agree with an LH-OA hybrid account on two important points. First, it embraces the conclusion that normative considerations do not entail metaphysical nonnaturalism (Broadbent 2019; Stempsey 2006). Although I will not rehearse the steps of that argument again here (see Section 1), it is vital to show exactly how Externalist SNC trades on this point. This version of Normativism views social evaluation as an objective fact about whether society has observed a particular procedure. This sociological fact ("social classification") is taken as being epistemically on par with the biophysical facts ("biological classification") that ground a disease condition. Sociology, a social science bound by methodological naturalism, empirically verifies whether the procedure of evaluation has taken place. Therefore, naturalism is not violated by the inclusion of normative considerations. Externalist SNC nevertheless self-identifies as a purely Normativist position, not as a hybrid view. It manages this feat by giving precedence to the social facts rather than the biological facts. The contingent social fact that we happen to judge an individual's condition the way we do (i.e., "diseased" vs "normal") after deliberation is allegedly more fundamental (metaphysically) because it establishes the antecedent background conditions against which the biological grounding relation holds necessarily if at all. That there was social deliberation of the right sort is consequently more important for these Normativists than the outcome of deliberation. The second point of agreement between Externalist SNC and the LH-OA hybrid account is their mutual acceptance of the best that the biological sciences have to offer. Externalist SNC's grounding relation can identify the biological basis for particular diseases and disorders in precisely the way that the LH-OA account's revised dysfunction condition suggests we should. Unlike its Internalist counterpart, it is not then beholden to the examination of arbitrarily delimited historical episodes or unwarranted emphasis on superseded medical science.

Considering these substantial points of overlap with the LH-OA hybrid account, Externalist SNC appears to be an eminently sensible form of Normativism. Why, then, is it not preferable to the hybrid account proposed herein?

Recall that Externalist SNC likens medicine to engineering in the sense that both aim to change the world rather than provide accurate description. Societies at large presumably establish the ends to which both medicine and engineering aim. Normativists in general have never been shy about making this point. Quill R. Kukla, for example, has argued for the priority of "social justice projects," whereupon "an understanding of health and disease is a part of a specific type of *normative* project – namely, that of determining the role that health *should* play in a larger theory of social justice" (2014, p. 516, italics in original). Sean Valles (2018) similarly argues for a "population health framework" that "seeks to decenter medicine and healthcare in the overall pursuit of health." According to him, "Most of the problems and most promising solutions to ill health lay outside the scope of biomedicine (safer workplaces, an end to racist housing discrimination, neighborhoods where people can safely walk, socialize and play, etc.)" (Valles 2018). At face value, there is nothing objectionable in these goals. Most would seemingly welcome their realization.

But whether these corrective/preventive measures (social "harm reduction" policies, institutional rearrangements) in fact constitute the "*most* promising" ways of dealing with "*most* of the problems of ill health" is, to put it mildly, debatable. In order to determine this, we must first grasp the biological details of individual and population-level functioning that LH-OA stresses. We must understand the biological mechanisms that causally mediate the influence of extrinsic (environmental) factors accompanying social inequity. It is crucially important to understand how and why social inequities manage to cause the adverse health consequences that they do. Only with this knowledge in hand can we be confident that the projects proposed in the interests of "social justice" will reliably deliver better health outcomes for all. Consider, for instance, how vital it is to understand the evolved life history trade-offs that an impoverished individual will exhibit when faced with severe caloric restrictions or a diet of diminished nutritional quality. Predictive adaptive response theory suggests that babies born to mothers who experience such conditions during pregnancy will likely develop a so-called "thrifty phenotype," which better prepares them for a life of deprivation (Bateson et al. 2004). With this information in hand, we can see why some seemingly self-evident ways of restoring equity would not in fact lend to social justice. Too hastily providing a generation of "thrifty" offspring with a diet containing more calories from sugars and fats has well-confirmed adverse effects, such as a dramatic increase in metabolic syndrome. Similar reasoning was applied to decide on the proper treatment of apparent anemia in cases of high-altitude hypoxia among Tibetans/Nepalese highlanders. Effective control requires comprehensive understanding along the lines first suggested by Tinbergen's framework.

The precedence Externalist SNC gives to the procedure of social evaluation seems unwarranted. It grants metaphysical priority to social deliberation, which is construed as an empirically verifiable fact and prerequisite ("anchoring fact") for subsequent medical evaluation. While this may make for a logically coherent form of naturalism, doing so comes with a cost. For it remains unclear why this sociologically established fact should be considered any more important to disease determination than other types of empirical (e.g., evolutionary and physiological) fact. The constitutive grounding basis for disease is, if anything, disjunctive by default, not hierarchical. Priority, whether it be of the metaphysical or evaluative variety, requires additional arguments to motivate it. Why should those of us who willing align our medical valuations with the modern biomedical establishment, which happen to be based largely if not exclusively on biophysical facts, cede normative authority to *nonmedical* social deliberations? Although this Externalist version of Normativism might like to feign otherwise, it must ultimately concede that social evaluative judgment alone can suffice for disease determination. From this perspective, evaluative deliberation of the sort involved in disease attribution finds its home among the means for realizing social justice, well-being, or some other conception of the common good. Insofar as medicine is considered just one among many such institutional "tools of construction," the biological conditions that a society at large classifies as "diseased/disordered" need not necessarily correspond with the recommendations of biomedical researchers.

The proposed analogy of medicine with engineering becomes ambiguous on this view. Both may be considered analogous insofar as they are prescriptive enterprises. But they have notably distinct aims. Surgeons/physicians, as David Gorski's remark in Section 3 makes clear, aim to restore or maintain *an existing conception of health*, one that is perhaps best conceived of as corresponding to guidelines established by Evolutionary Medicine and LH-OA's more comprehensive elaboration of (dys)function. Externalist SNC, in contrast, thinks of medicine as a means of changing the health standards according to what a society at large *would like them to be*. While this Normativist account can take the best biomedicine of the day on board, which it may well be inclined to do in many cases, it can easily choose not to on the grounds that there is no corresponding socially agreed upon "disease" with such biophysical grounding conditions.

The hybrid LH-OA account, in contrast, explicitly denies this type of autonomy. Disabilities present a good example. By almost any *biological* standard, hereditary deafness is considered a disorder as opposed to a mere (harmless) variation. The biological conditions that ground hereditary deafness are well known. Externalist SNC strikes many as capricious in that it can justify the classification of hereditary deafness as a disorder at one point in time, while revoking this classification at another. It might characterize deafness as a disorder in order to justify the allocation of further funding and medical research into its biological basis. One could argue that this depiction has borne fruits in the form of technological advances (cochlear implants) and successful candidate gene analyses (Chen et al. 2021), which, in the case of the former, have helped to ameliorate some of the disadvantages associated with deafness. However, Externalist SNC could just as easily justify the classification of hereditary deafness as normal, which would then coincide with the way that some in the deaf community currently see their condition (i.e., as a linguistic minority suffering discrimination). Some in the deaf community might find this alternative depiction of deafness as non-disordered appealing if gene-editing technologies were to enable the elimination of genetic mutations (e.g., *GAS2*) implicated in hereditary deafness. Either way, it is important to notice that the plausibility of any such reversal in medical status is anchored in the state of the science. The transition from disordered to well ordered in this example would involve allocating resources to the replacement of a *detrimental* mutational variant with a "normal" variant or some other form of *corrective* technological supplementation. The possibility and availability of such interventions does not imply that hereditary deafness or the mutations responsible for it are no longer pathological.

While LH-OA resists the precedence that Externalist SNC grants to evaluative judgment, it cannot expel value considerations entirely. Nor should it. The "harm component" is what importantly distinguishes any hybrid account from pure Naturalism (ONR). Setting aside the precedence Normativists give to social deliberation, I share their conviction that something closely resembling the set of four normative conditions from social deliberation (discussed earlier, see Conley and Glackin 2021, p. 10) is indeed necessary for disease determination. This set of conditions is, I would argue, better understood as a sophisticated attempt at elaborating the somewhat vague harm condition from Wakefield's original proposal. In this limited sense, there is a great deal of agreement when it comes to the role of values in determining disease.

While the harm condition for hybrid LH-OA might be closely aligned with Conley and Glackin's (2021) conception of social deliberation, it would differ in at least one respect. The primary difference regards how to interpret the intolerability condition (condition (1) from above). To claim that some condition is harmful implies a judgment to the effect that it is intolerable. The distinguishing normative feature of the hybrid LH-OA account is that intolerability remains a brute judgment of the afflicted individual. It is not open to negotiation. Proponents of Externalist SNC, in contrast, seem to believe that evaluations of intolerability are *social* judgments just like the determinations of disease in which they constitutively figure. There is, for them, a matter of fact about whether a condition is intolerable, and individuals can accordingly err in their judgments. Externalist SNC must hold to this if it is to avoid pernicious forms of relativism (in this case radical (inter-)subjectivism). An individual's assessment of tolerability may well coincide with consensus opinion in many if not most cases. However, in some cases it may not. Patients with disabilities, for instance, typically rate their conditions less severe than outsiders do. That a social consensus or "majority rule" is the best and seemingly only hope for thwarting counterintuitive, subjective assessments of (in)tolerability consequently appears to be a weakness of the view. Majorities can, after all, be tyrannical.

The LH-OA hybrid account must likewise summon the resources to reject the immediate implication of harm from an individual's claim of intolerability. Instead of turning directly to social consensus, as Externalist SNC does, it would attempt to limit individual evaluations of tolerability by giving greater deliberative weight to biophysical grounding conditions (i.e., the dysfunction condition). What is, in effect, being questioned is the purported necessity of any social (nonmedical) judgment that supports an individual's assessment of tolerability. Individuals who do *not* realize at least one of the four "ways of going wrong" adopted by LH-OA, but who still claim their condition intolerable, could accordingly be said to experience something that is, by their very own standards, genuinely intolerable. That individual judgment would hold irrespective of whether or not society at large condones it. Either way, the form of "intolerability" then at issue, from a medical standpoint, is only of the kind that would also, for example, be attributable to one who must reside with an inamicably divorced partner. It is a serious inconvenience, but not a harm befitting medical attention. Scientific consensus, rather than societal consensus at large, is the preferred bulwark against radical subjectivism when it comes to harms of the medical variety. Whether it be the evaluation of a single individual, a minority, or even a majority, evaluations of intolerability *alone* never suffice to determine medically relevant harm or pathology. Disease determination always requires some sort of biological basis.

5.4 Conclusion

Sophisticated Normativist accounts of disease or disorder have blind spots. This holds true for even the most thoroughly worked through version of Normativism, what I have called "Externalist SNC." A sophisticated, multi-faceted view informed by modern evolutionary theorizing – the LH-OA hybrid account – has some compelling strengths. I have attempted to state these strengths clearly. While I endorse the hybrid LH-OA, it must be admitted that it comes with pitfalls that have yet to be adequately addressed. Furthermore, it remains to be seen whether focusing on the normative issue of (in)tolerance constitutes a noteworthy improvement. That I have not addressed or resolved all of these difficulties may consequently leave some readers dissatisfied. To those who feel this way, I hasten to point out three facts. First, philosophical discourse rarely results in unconditional surrender or outright victory. Second, the cardinal aim of this introductory text was to make evident the shortcomings of the prevailing dialectic (Normativism vs. Naturalism; SNC vs. ONR) in the philosophy of medicine. The roadmap for a more focused and fruitful exchange between the LH-OA hybrid account and Externalist SNR is now on offer. Both can accept the best that the biosciences have to offer, and each recognizes that value considerations in some form or other are ineliminable. There nevertheless remain significant differences concerning how best to characterize the role of (non-epistemic) values and the connection these have to findings in the biomedical sciences. Third, the LH-OA hybrid account I sketch diverges significantly from the consensus in the literature. It might thus be considered somewhat "original" and interesting for those who follow the relevant debates in the philosophy of medicine. At least, this is my hope. What is perhaps most obvious is that a great deal of work remains to be done in this area. If readers take nothing else away from this short treatise, they should recognize that such work must be theoretical and empirical as well as conceptual in nature.

References

Alcock, J., & Schwartz, M. D. (2011). A clinical perspective in evolutionary medicine: What we wish we had learned in medical school. *Evolution: Education and Outreach, 4*(4), 574–579.

American Psychiatric Association. (1952). *Diagnostic and statistical manual of mental disorders: DSM-I* (1st ed.). American Psychiatric Association.

Amundson, R. (2000). Against normal function. *Studies in History and Philosophy of Science Part C: Studies in History and Philosophy of Biological and Biomedical Sciences, 31*(1), 33–53.

Amundson, R., & Lauder, G. V. (1994). Function without purpose. *Biology and Philosophy, 9*(4), 443–469.

Barkow, J. H., Cosmides, L. E., & Tooby, J. E. (1992). *The adapted mind: Evolutionary psychology and the generation of culture*. Oxford University Press.

Bateson, P., Barker, D., Clutton-Brock, T., Deb, D., D'Udine, B., Foley, R. A., Gluckman, P., Godfrey, K., Kirkwood, T., Lahr, M. M., McNamara, J., Metcalfe, N. B., Monaghan, P., Spencer, H. G., & Sultan, S. E. (2004). Developmental plasticity and human health. *Nature, 430*(6998), 419–421. https://doi.org/10.1038/nature02725

Bateson, P., & Laland, K. N. (2013). Tinbergen's four questions: An appreciation and an update. *Trends in Ecology & Evolution, 28*(12), 712–718.

Beall, C. M. (2014). Adaptation to high altitude: Phenotypes and genotypes. *Annual Review of Anthropology, 43*(1), 251–272. https://doi.org/10.1146/annurev-anthro-102313-030000

Beall, C. M. (2019). Biodiversity of human populations in mountain environments. In Ch. Körner and E. M. Spehn (Ed.), *Mountain biodiversity: A global assessment* (pp. 199–210). Routledge.

Beall, C. M., Cavalleri, G. L., Deng, L., Elston, R. C., Gao, Y., Knight, J., Li, C., Li, J. C., Liang, Y., McCormack, M., Montgomery, H. E., Pan, H., Robbins, P. A., Shianna, K. V., Tam S. C., Tsering, N., Veeramah, K. R., Wang, W., Wangdui, P., Weale, M. E., Xu, Y., Xu, Z., Yang, L., Zaman, M. J., Zeng, C., Zhang, L., Zhang, X., Zhaxi, P., & Zheng, Y. T. (2010). Natural selection on EPAS1 (HIF2α) associated with low hemoglobin concentration in Tibetan highlanders. *Proceedings of the National Academy of Sciences, 107*(25), 11459–11464.

References

Bechtel, W., & Richardson, R. C. (1993). *Discovering complexity: Decomposition and localization as strategies in scientific research*. Princeton University Press.

Boorse, C. (1977). Health as a theoretical concept. *Philosophy of Science, 44*, 542–574.

Boorse, C. (1997). A rebuttal on health. In J. M. Humber and R. F. Almeder (Ed.), *What is disease?* (pp. 1–134). Humana Press.

Boorse, C. (2002). A rebuttal on functions. In A. Ariew, R. C. Cummins, & M. Perlman (Eds.), *Functions: New essays in the philosophy of psychology and biology* (p. 63). Oxford University Press.

Boorse, C. (2014). A second rebuttal on health. *Journal of Medicine and Philosophy, 39*(6), 683–724. https://doi.org/10.1093/jmp/jhu035

Bourrat, P., & Griffiths, P. E. (2021). The idea of mismatch in evolutionary medicine. *British Journal for the Philosophy of Science*. https://doi.org/10.1086/716543

Bouverot, P. (2012). *Adaptation to altitude-hypoxia in vertebrates* (Vol. 16). Springer Science & Business Media.

Bowlby, J. (1969). *Attachment and loss: Vol. 1: Attachment*. The Hogarth Press and the Institute of Psycho-Analysis. www.pep-web.org.ezproxy1.library.usyd.edu.au/document.php?id=IPL.079.0001A

Boyd, R., & Richerson, P. J. (1988). *Culture and the evolutionary process*. University of Chicago Press.

Broadbent, A. (2019). Health as a secondary property. *The British Journal for the Philosophy of Science, 70*(2), 609–627.

Brown, J. S., & Venable, D. L. (1986). Evolutionary ecology of seed-bank annuals in temporally varying environments. *The American Naturalist, 127*(1), 31–47.

Brown, P. (1990). The name game: Toward a sociology of diagnosis. *The Journal of Mind and Behavior, 11*(3–4), 385–406.

Buklijas, T., & Gluckman, P. (2013). From evolution and medicine to evolutionary medicine. In M. Ruse (Ed.), *The Cambridge encyclopedia of Darwin and evolutionary thought* (pp. 505–513). Cambridge University Press.

Camperio-Ciani, A., Corna, F., & Capiluppi, C. (2004). Evidence for maternally inherited factors favouring male homosexuality and promoting female fecundity. *Proceedings of the Royal Society of London. Series B: Biological Sciences, 271*(1554), 2217–2221.

Canguilhem, G. (2012). *On the normal and the pathological* (Vol. 3). Springer Science & Business Media.

Carnap, R. (1950). *Logical foundations of probability*. University of Chicago Press.

Chen, T., Rohacek, A. M., Caporizzo, M., Nankali, A., Smits, J. J., Oostrik, J., Lanting, C. P., Kücük, E., Gilissen, C., & van de Kamp, J. M. (2021). Cochlear supporting cells require GAS2 for cytoskeletal architecture and hearing. *Developmental Cell*, *56*(10), 1526–1540. e7.

Cho, J. I., Basnyat, B., Jeong, C., Di Rienzo, A., Childs, G., Craig, S. R., Sun, J., & Beall, C. M. (2017). Ethnically Tibetan women in Nepal with low hemoglobin concentration have better reproductive outcomes. *Evolution, Medicine, and Public Health*, *2017*(1), 82–96.

Christie, J. R., Brusse, C., Bourrat, P., Takacs, P., & Griffiths, P. E. (2022). Do proper functions explain the existence of traits? *Australasian Philosophical Review* *6*(4), 335–359.

Conley, B. A., & Glackin, S. N. (2021). How to be a naturalist and a social constructivist about diseases. *Philosophy of Medicine*, *2*(1), 1–21.

Cooper, R. (2002). Disease. *Studies in History and Philosophy of Science Part C: Studies in History and Philosophy of Biological and Biomedical Sciences*, *33*(2), 263–282.

Cosmi, E., Fanelli, T., Visentin, S., Trevisanuto, D., & Zanardo, V. (2011). Consequences in infants that were intrauterine growth restricted. *Journal of Pregnancy*, *2011*(1), 364381. https://doi.org/10.1155/2011/364381

Cummins, R. (1975). Functional analysis. *Journal of Philosophy*, *72*(November), 741–764.

Darwin, C. (1964). *On the origin of species: A facsimile of the first edition*. Harvard University Press.

Darwin, C., Wallace, A. R., Lyell, S. C., & Hooker, J. D. (1858). On the tendency of species to form varieties: And on the perpetuation of varieties and species by natural means of selection. *Zoological Journal of the Linnean Society*, *3*(9), 46–62.

Darwin, E. (1809). Zoonomia, Or, The Laws of Organic Life: In Three Parts (Issue 17339). Thomas & Andrews, JT Buckingham, printer.

Davies, N. B., & Krebs, J. R. (1984). *Behavioural ecology*. Blackwell Scientific.

Diep, F. (2017). Why do some doctors reject evolution? Pacific Standard. https://psmag.com/social-justice/how-ben-how

Dobzhansky, T. (1973). Nothing in biology makes sense except in the light of evolution. *The American Biology Teacher*, *35*, 125–129.

Downie, J. R. (2004). Evolution in health and disease: The role of evolutionary biology in the medical curriculum. *Bioscience Education*, *4*(1), 1–18.

Engelhardt Jr, H. T. (1976). Ideology and etiology. *The Journal of Medicine and Philosophy*, *1*(3), 256–268.

Ereshefsky, M. (2009). Defining "health" and "disease." *Studies in History and Philosophy of Science Part C: Studies in History and Philosophy of Biological and Biomedical Sciences*, *40*(3), 221–227.

Ewald, P. W. (1994). *Evolution of infectious disease*. Oxford University Press on Demand.

Flint, J., Harding, R. M., Boyce, A. J., & Clegg, J. B. (1998). The population genetics of the haemoglobinopathies. *Baillieres Clinical Haematology*, *11*(1), 1–51.

Foucault, M. (1973). *The birth of the clinic: An archaeology of medical perception*, trans. AM Sheridan Smith. Pantheon.

Framingham Heart Study. www.framinghamheartstudy.org/

Garson, J. (2019). *What biological functions are and why they matter*. Cambridge University Press.

Garson, J. (2021). The developmental plasticity challenge to Wakefield's view. In *Defining mental disorder: Jerome Wakefield and his critics*. (pp. 335–351). MIT Press.

Gassmann, M., Mairbäurl, H., Livshits, L., Seide, S., Hackbusch, M., Malczyk, M., Kraut, S., Gassmann, N. N., Weissmann, N., & Muckenthaler, M. U. (2019). The increase in hemoglobin concentration with altitude varies among human populations. *Annals of the New York Academy of Sciences*, *1450*(1), 204–220.

Genereux, D. P., & Bergstrom, C. T. (2005). Evolution in action: Understanding antibiotic resistance. Evolutionary Science and Society: Educating a New Generation. AIBS/BCSC, Washington, DC, 145–153.

Glackin, S. N. (2010). Tolerance and illness: The politics of medical and psychiatric classification. *Journal of Medicine and Philosophy*, *35*(4), 449–465. https://doi.org/10.1093/jmp/jhq035

Glackin, S. N. (2019). Grounded disease: Constructing the social from the biological in medicine. *The Philosophical Quarterly*, *69*(275), 258–276.

Gluckman, P. D., Beedle, A. S., Buklijas, T., Low, F., & Hanson, M. A. (2016). *Principles of evolutionary medicine* (2nd ed.). Oxford University Press.

Gluckman, P. D., Hanson, M. A., Spencer, H. G., & Bateson, P. (2005). Environmental influences during development and their later consequences for health and disease: Implications for the interpretation of empirical studies. *Proceedings of the Royal Society B-Biological Sciences*, *272*(1564), 671–677. https://doi.org/10.1098/rspb.2004.3001

Godfrey-Smith, P. (1993). Functions: Consensus without unity. *Pacific Philosophical Quarterly, 74*(3), 196–208.
Godfrey-Smith, P. (1994). A modern history theory of functions. *Noûs, 28*(3), 344–362. https://doi.org/10.2307/2216063
Godfrey-Smith, P. (2001). Three kinds of adaptationism. In S. Orzack & E. Sober (Eds.), *Optimality and adaptation* (pp. 335–357). Cambridge University Press.
Goosens, W. K. (1980). Values, health, and medicine. *Philosophy of Science, 47*(1), 100–115.
Gould, S. J., & Lewontin, R. C. (1979). The spandrels of San Marco and the Panglossian paradigm: A critique of the adaptationist programme. *Proceedings of the Royal Society of London. Series B. Biological Sciences, 205*(1161), 581–598.
Griffiths, P. E. (1993). Functional analysis and proper functions. *British Journal for Philosophy of Science, 44*(3), 409–422. https://doi.org/10/ddkn3x
Griffiths, P. E., & Matthewson, J. (2018). Evolution, dysfunction, and disease: A reappraisal. *The British Journal for the Philosophy of Science, 69*(2), 301–327.
Haig, D. (2015). Maternal–fetal conflict, genomic imprinting and mammalian vulnerabilities to cancer. *Philosophical Transactions of the Royal Society B: Biological Sciences, 370*(1673), 20140178.
Harper, C. V., Woodcock, D. J., Lam, C., Garcia-Albornoz, M., Adamson, A., Ashall, L., Rowe, W., Downton, P., Schmidt, L., West, S., Spiller, D. G., Rand, D. A., & White, M. R. H. (2018). Temperature regulates NF-κB dynamics and functionthrough timing of A20 transcription. *Proceedings of the National Academy Science USA, 115*(22), E5243–E5249.
Haselton, M. G., & Buss, D. M. (2000). Error management theory: A new perspective on biases in cross-sex mind reading. *Journal of Personality and Social Psychology, 78*(1), 81.
Iemmola, F., & Camperio-Ciani, A. (2009). New evidence of genetic factors influencing sexual orientation in men: Female fecundity increase in the maternal line. *Archives of Sexual Behavior, 38*(3), 393–399.
Kingma, E. (2010). Paracetamol, poison, and polio: Why Boorse's account of function fails to distinguish health and disease. *British Journal for the Philosophy of Science, 61*(2), 241–264.
Kitcher, P. (1997). *The lives to come*. Simon & Schuster.
Klinger, K. W. (1994). Genetics of cystic fibrosis. *Seminars in Respiratory Medicine, 6*, 243.

Koirala, S., Nakano, M., Arima, H., Takeuchi, S., Ichikawa, T., Nishimura, T., Ito, H., Pandey, B. D., Pandey, K., & Wada, T. (2018). Current health status and its risk factors of the Tsarang villagers living at high altitude in the Mustang district of Nepal. *Journal of Physiological Anthropology, 37*(1), 1–10.

Kuhn, T. S. (2011). *The essential tension*. University of Chicago Press.

Kukla, R. (2014). Medicalization, "normal function," and the definition of health. In *The Routledge companion to bioethics* (pp. 539–554). Routledge.

Laland, K. N., Uller, T., Feldman, M. W., Sterelny, K., Müller, G. B., Moczek, A., Jablonka, E., & Odling-Smee, J. (2015). The extended evolutionary synthesis: Its structure, assumptions and predictions. *Proceedings of the Royal Society B: Biological Sciences, 282*(1813), 20151019.

Lemoine, M. (2013). Defining disease beyond conceptual analysis: An analysis of conceptual analysis in philosophy of medicine. *Theoretical Medicine and Bioethics, 34*: 309–325. https://doi.org/10.1007/s11017-013-9261-5

Lewens, T. (2015). *The biological foundations of bioethics*. OUP Oxford.

Margolis, J. (1976). The concept of disease. *Journal of Medicine and Philosophy, 1*(3), 238–255

Matthewson, J., & Griffiths, P. E. (2017). Biological criteria of disease: Four ways of going wrong. *The Journal of Medicine and Philosophy: A Forum for Bioethics and Philosophy of Medicine, 42*(4), 447–466. https://doi.org/10/gbtpb2

Mayr, E. (1961). Cause and effect in biology. *Science, 134*(3489), 1501–1506.

McMullin, E. (1982). Values in science. *PSA: Proceedings of the Biennial Meeting of the Philosophy of Science Association, 1982*(2), 3–28.

Méthot, P.-O. (2011). Research traditions and evolutionary explanations in medicine. *Theoretical Medicine and Bioethics, 32*(1), 75–90. https://doi.org/10.1007/s11017-010-9167-4

Mill, J. S. (1865). A System of Logic, 2 vols. (London, 1846). Principles of Political Economy, 2.

Millikan, R. G. (1984). *Language, thought, and other biological categories: New foundations for realism*. MIT Press.

Montévil, M., & Mossio, M. (2015). Biological organisation as closure of constraints. *Journal of Theoretical Biology, 372*, 179–191.

Mossio, M., & Bich, L. (2017). What makes biological organisation teleological? *Synthese, 194*(4), 1089–1114.

Mossio, M., Saborido, C., & Moreno, A. (2009). An organizational account of biological functions. *The British Journal for the Philosophy of Science, 60*(4), 813–841.

Mowat, A. (2017). "Why does cystic fibrosis display the prevalence and distribution observed in human populations?" *Current Pediatric Research, 21*(1), 164–171.

Murphy, D. (2006). *Psychiatry in the scientific image*. MIT Press.

Murphy, D. (2015). Concepts of disease and health. In E. N. Zalta (Ed.), *The Stanford encyclopedia of philosophy* (Spring 2015). http://plato.stanford.edu/archives/spr2015/entries/health-disease/

Murphy, T. F. (1997). *Gay science: The ethics of sexual orientation research*. Columbia University Press.

Neander, K. L. (1983). *Abnormal psychobiology: A thesis on the "anti-psychiatry Debate" and the relationship between psychology and biology*. La Trobe University.

Neander, K. L. (1991a). Functions as selected effects: The conceptual analyst's defense. *Philosophy of Science, 58*, 168–184.

Neander, K. L. (1991b). The teleological notion of "function." *Australasian Journal of Philosophy, 69*(4), 454–468.

Neander, K. L. (2017). *A mark of the mental: In defense of informational teleosemantics*. MIT Press.

Nesse, R. M. (2001). The smoke detector principle. Natural selection and the regulation of defensive responses. *Annals of the New York Academy of Sciences, 935*, 75–85.

Nesse, R. M. (2013). Tinbergen's four questions, organized: A response to Bateson and Laland. *Trends in Ecology & Evolution, 28*(12), 681–682.

Nesse, R. M. (2019). *Good reasons for bad feelings: Insights from the frontier of evolutionary psychiatry*. Penguin.

Nesse, R. M., & Schiffman, J. D. (2003). Evolutionary biology in the medical school curriculum. *BioScience, 53*(6), 585–587.

Nesse, R. M., & Williams, G. C. (1994). *Why we get sick: The new science of Darwinian medicine*. Times Books.

Nesse, R. M., & Williams, G. C. (1998). Evolution and the origins of disease. *Scientific American, 279*(5), 86–93.

Okumiya, K., Sakamoto, R., Ishimoto, Y., Kimura, Y., Fukutomi, E., Ishikawa, M., Suwa, K., Imai, H., Chen, W., & Kato, E. (2016). Glucose intolerance associated with hypoxia in people living at high altitudes in the Tibetan highland. *BMJ Open, 6*(2), e009728.

Perlman, R. L. (2013). *Evolution and medicine*. Oxford University Press.

Petousi, N., & Robbins, P. A. (2014). Human adaptation to the hypoxia of high altitude: The Tibetan paradigm from the pregenomic to the postgenomic era. *Journal of Applied Physiology, 116*(7), 875–884.

Pigliucci, M., & Müller, G. B. (2010). *Evolution–the extended synthesis*. MIT Press.

Pittendrigh, C. S. (1958). Adaptation, natural selection, and behavior. *Behavior and Evolution, 390*, 416.

Rabinowitz, P. M., Natterson-Horowitz, B. J., Kahn, L. H., Kock, R., & Pappaioanou, M. (2017). Incorporating one health into medical education. *BMC Medical Education, 17*, 45.

Ramsey, G., & Aaby, B. H. (2022). The proximate-ultimate distinction and the active role of the organism in evolution. *Biology & Philosophy, 37*, 31.

Regitz-Zagrosek, V., Roos-Hesselink, J. W., Bauersachs, J., Blomström-Lundqvist, C., Cifkova, R., De Bonis, M., Iung, B., Johnson, M. R., Kintscher, U., & Kranke, P. (2018). 2018 ESC guidelines for the management of cardiovascular diseases during pregnancy: The task force for the management of cardiovascular diseases during pregnancy of the European Society of Cardiology (ESC). *European Heart Journal, 39*(34), 3165–3241.

Riise, H. K. R., Sulo, G., Tell, G. S., Igland, J., Nygård, O., Iversen, A. C., & Daltveit, A. K. (2018). Association between gestational hypertension and risk of cardiovascular disease among 617, 589 Norwegian women. *Journal of the American Heart Association, 7*(10), e008337. https://doi.org/10.1161/JAHA.117.008337

Rohlf, F. J., & Sokal, R. R. (1995). *Biometry: The principles and practice of statistics in biological research*. W.H. Freeman and Company.

Roughgarden, J. (2012a). Teamwork, pleasure and bargaining in animal social behaviour. *Journal of Evolutionary Biology, 25*(7), 1454–1462.

Roughgarden, J. (2012b). The social selection alternative to sexual selection. *Philosophical Transactions of the Royal Society B: Biological Sciences, 367*(1600), 2294–2303.

Ruse, M. (1988). *Homosexuality: A philosophical inquiry*. Blackwell.

Ruse, M. (1996). *Monad to man: The concept of progress in evolutionary biology*. Harvard University Press.

Saborido, C., & Moreno, A. (2015). Biological pathology from an organizational perspective. *Theoretical Medicine and Bioethics, 36*(1), 83–95.

Sarna, K., Brittenham, G. M., & Beall, C. M. (2020). Detecting anaemia at high altitude. *Evolution, Medicine, and Public Health, 2020*(1), 68–69.

Schaffner, K. F. (1993). *Discovery and explanation in biology and medicine*. University of Chicago Press.

Schwartz, P. H. (2007). Defining dysfunction: Natural selection, design, and drawing a line. *Philosophy of Science, 74*(3): 364–385.

Sedgwick, P. (1982). *Psycho politics: Laing, Foucault, Goffman, Szasz, and the future of mass psychiatry*. Harper & Row.

Segerstrale, U. (2000). *Defenders of the truth: The battle for science in the sociobiology debate and beyond*. Oxford University Press.

Shafer-Landau, R. (2003). *Moral realism: A defence*. Oxford University Press on Demand.

Shea, N. (2013). Naturalising representational content. *Philosophy Compass, 8*(5), 496–509.

Simon, J. R., Carel, H., & Bird, A. (2017). Understanding disease and illness. *Theoretical Medicine and Bioethics, 38*(4): 239–244.

Slee, V. N. (1978). *The International classification of diseases: Ninth revision (ICD-9)*. American College of Physicians.

Sociobiology Study Group of Science for the People. (1976). Dialogue. The critique: Sociobiology: Another biological determinism. *BioScience, 1*(March), 182–186.

Stearns, S. C. (1992). *The evolution of life histories*. Oxford University Press.

Stearns, S. C., & Medzhitov, R. (2015). *Evolutionary medicine*. Sinauer Associates, Inc..

Stegenga, J. (2018). *Medical nihilism*. Oxford University Press.

Stempsey, W. E. (2006). *Disease and diagnosis: Value-dependent realism* (Vol. 63). Springer Science & Business Media.

Taylor, R., & Rieger, A. (1985). Medicine as social science: Rudolf Virchow on the typhus epidemic in Upper Silesia. *International Journal of Health Services, 15*(4), 547–559.

Thagard, P. (2018). *How scientists explain disease*. Princeton University Press.

Tinbergen, N. (1963). On aims and methods of ethology. *Zeitschrift Für Tierpsychologie, 20*(4), 410–433.

Tremblay, J. C., & Ainslie, P. N. (2021). Global and country-level estimates of human population at high altitude. *Proceedings of the National Academy of Sciences*, 118(18), e2102463118.

Valles, S. A. (2012). Evolutionary medicine at twenty: Rethinking adaptationism and disease. *Biology & Philosophy, 27*(2), 241–261.

Valles, S. A. (2018). *Philosophy of population health: Philosophy for a new public health era*. Routledge.

Virchow, V. (1858). Cellular pathology is based on physiological and pathological tissue theory. Twenty lectures held during the months of February, March and April at the Pathological Institute in Berlin. A. Hirschwald; Library of Congress. www.loc.gov/item/06041231/

Wachbroit, R. (1994). Normality as a biological concept. *Philosophy of Science, 61*(4), 579–591.

Wagner, P. D., Simonson, T. S., Wei, G., Wagner, H. E., Wuren, T., Qin, G., Yan, M., & Ge, R. L. (2015). Sea-level haemoglobin concentration is associated with greater exercise capacity in Tibetan males at 4200 m. *Experimental Physiology, 100*(11), 1256–1262.

Wakefield, J. C. (1992a). Disorder as harmful dysfunction: A conceptual critique of DSM-III-R's definition of mental disorder. *Psychological Review, 99*(2), 232.

Wakefield, J. C. (1992b). The concept of mental disorder: On the boundary between biological facts and social values. *American Psychologist, 47*(3), 373.

Wang, E. E., Einarson, T. R., Kellner, J. D., & Conly, J. M. (1999). Antibiotic prescribing for Canadian preschool children: Evidence of overprescribing for viral respiratory infections. *Clinical Infectious Diseases, 29*(1), 155–160.

Wells, J. C. K. (2012). A critical appraisal of the predictive adaptive response hypothesis. *International Journal of Epidemiology, 41*(1), 229–235. https://doi.org/10.1093/ije/dyr239

Wexler, N. S., Collett, L., & Wexler, A. R. et al. (2016). Incidence of adult Huntington's disease in the UK: A UK-based primary care study and a systematic review. *BMJ Open, 6*, e009070. https://doi.org/10.1136/bmjopen-2015-009070

Whelton, P. K., Williams, B., Qamar, A., Braunwald, E., Hwang, K. O., Thomas, E. J., Petersen, L. A., Sharaf, R. N., & Diamond, L. C. (2018). Hypertension as a global challenge. *Journal of the American Medical Association, 320*(17), 1721–1723.

Wieselthaler, G. M., Schima, H., Hiesmayr, M., Pacher, R., Laufer, G., Noon, G. P., DeBakey, M., & Wolner, E. (2000). First clinical experience with the DeBakey VAD continuous-axial-flow pump for bridge to transplantation. *Circulation, 101*(4), 356–359.

Williams, G. C., & Nesse, R. M. (1991). The Dawn of Darwinian medicine. *The Quarterly Review of Biology, 66*(1), 1–22.

Wilson, E. O. (1975). *Sociobiology: The new synthesis*. Harvard University Press.

Withrock, I. C., Anderson, S. J., Jefferson, M. A., McCormack, G. R., Mlynarczyk, G. S., Nakama, A., Lange, J. K., Berg, C. A., Acharya, S., & Stock, M. L. (2015). Genetic diseases conferring resistance to infectious diseases. *Genes & Diseases, 2*(3), 247–254.

Woodward, J. (2003). *Making things happen: A theory of causal explanation*. Oxford University Press.

World Health Organization. (2017). Technical meeting: Use and interpretation of haemoglobin concentrations for assessing anaemia status in individuals and populations. www.who.int/nutrition/events/2017-meeting-haemoglobin-concentrations-anaemia-29novto1dec/en/

Wright, L. (1973). Functions. *Philosophical Review, 82*(2), 139–168.

Zampieri, F. (2009). Medicine, evolution, and natural selection: An historical overview. *The Quarterly Review of Biology, 84*(4), 333–355.

Acknowledgments

I am grateful to the editors of this series, Grant Ramsey and the late Michael Ruse. Without their encouragement, advice, and most of all understanding, this Element would never have seen the light of day. I would also like to thank two anonymous reviewers for Cambridge University Press. Their diligence and constructive criticism saved me from countless (sometimes silly) errors. As no book is written in isolation, my thinking has been shaped in many ways by interactions with colleagues at The University of Sydney and Macquarie University as well as fellow members of the International Society for Evolution, Medicine, and Public Health. Regular exchanges with Paul Griffiths, Randolph Nesse, Dominic Murphy, and the Theory and Method in Biosciences working group were particularly influential. I am plainly in debt to my wife Shannon for just about everything.

I gratefully acknowledge the financial support of the John Templeton Foundation (grant numbers 62220 and 63452). The opinions expressed in this Element are my own and not those of the John Templeton Foundation. This research was also supported under Australian Research Council's Discovery Projects funding scheme (project number FL170100160).

Cambridge Elements ≡

Philosophy of Biology

Grant Ramsey
KU Leuven

Grant Ramsey is a BOFZAP research professor at the Institute of Philosophy, KU Leuven, Belgium. His work centers on philosophical problems at the foundation of evolutionary biology. He has been awarded the Popper Prize twice for his work in this area. He also publishes in the philosophy of animal behavior, human nature, and the moral emotions. He runs the Ramsey Lab (theramseylab.org), a highly collaborative research group focused on issues in the philosophy of the life sciences.

About the Series

This Cambridge Elements series provides concise and structured introductions to all of the central topics in the philosophy of biology. Contributors to the series are cutting-edge researchers who offer balanced, comprehensive coverage of multiple perspectives, while also developing new ideas and arguments from a unique viewpoint.

Cambridge Elements ≡

Philosophy of Biology

Elements in the Series

Cultural Selection
Tim Lewens

Biological Organization
Leonardo Bich

Controlled Experiments
Jutta Schickore

Slime Mould and Philosophy
Matthew Sims

Explanation in Biology
Lauren N. Ross

Philosophy of Physiology
Maël Lemoine

The Organism
Jan Baedke

Human Cognitive Diversity
Ingo Brigandt

Modelling Evolution
Walter Veit

The Scope of Evolutionary Thinking
Thomas A. C. Reydon

What Is Life? Revisited
Daniel J. Nicholson

Biology and Medical Theory
Peter Takacs

A full series listing is available at: www.cambridge.org/EPBY

Printed by Integrated Books International,
United States of America